Springer Tracts in Electrical and Electronics Engineering

Springer Tracts in Electrical and Electronics Engineering (STEEE) publishes the latest developments in Electrical and Electronics Engineering - quickly, informally and with high quality. The intent is to cover all the main branches of electrical and electronics engineering, both theoretical and applied, including:

- Signal, Speech and Image Processing
- Speech and Audio Processing
- Image Processing
- Human-Machine Interfaces
- Digital and Analog Signal Processing
- Microwaves, RF Engineering and Optical Communications
- Electronics and Microelectronics, Instrumentation
- Electronic Circuits and Systems
- Embedded Systems
- Electronics Design and Verification
- Cyber-Physical Systems
- Electrical Power Engineering
- Power Electronics
- Photovoltaics
- Energy Grids and Networks
- Electrical Machines
- Control, Robotics, Automation
- Robotic Engineering
- Mechatronics
- Control and Systems Theory
- Automation
- Communications Engineering, Networks
- Wireless and Mobile Communication
- Internet of Things
- Computer Networks

Within the scope of the series are monographs, professional books or graduate textbooks, edited volumes as well as outstanding PhD theses and books purposely devoted to support education in electrical and electronics engineering at graduate and post-graduate levels.

Review Process

The proposal for each volume is reviewed by the main editor and/or the advisory board. The books of this series are reviewed in a single blind peer review process.

Ethics Statement for this series can be found in the Springer standard guidelines here https://www.springer.com/us/authors-editors/journal-author/journal-author-helpdesk/before-you-start/before-you-start/1330#c14214.

Geethu Joseph · Chandra R. Murthy

Sparsity-Constrained Linear Dynamical Systems

From Compressed Sensing to Control Theory

Springer

Geethu Joseph
Signal Processing Systems Group
Delft University of Technology
Delft, The Netherlands

Chandra R. Murthy
Department of Electrical Communication Engineering
Indian Institute of Science Bangalore
Bengaluru, India

ISSN 2731-4200 ISSN 2731-4219 (electronic)
Springer Tracts in Electrical and Electronics Engineering
ISBN 978-981-97-7089-2 ISBN 978-981-97-7090-8 (eBook)
https://doi.org/10.1007/978-981-97-7090-8

This Springer imprint is published by the registered company Springer Nature Singapore Pte Ltd.
The registered company address is: 152 Beach Road, #21-01/04 Gateway East, Singapore 189721, Singapore

To our families
Jomon, Daniel, and Hannah
Divya and Anvi

Preface

At the intersection of control engineering and signal processing lies the upcoming field of *sparse* control and state estimation of linear dynamical systems. This field centers on linear dynamical systems with states or control inputs having a small number nonzero entries compared to their dimensions. This book serves as a comprehensive guide, offering an extensive overview, critical insights, and in-depth discussions about recent theoretical advancements and algorithmic developments within this nascent research area.

Sparsity constraints arise naturally in many real-world systems. Diffusion processes in complex networks like disease or epidemic propagation in human society, air or water pollution, and virus dissemination in computer/mobile phone networks, are example systems where the *state* of the system often exhibits sparse *initialization*. Likewise, imposing sparsity on *inputs* applied to the system becomes essential to select a small subset of available sensors or actuators at each time instant due to energy, bandwidth, or physical network limitations. For instance, in energy-constrained networks, employing energy-conscious scheduling of sensors or actuators through sparse inputs can significantly prolong network lifespan. Additionally, employing sparse control inputs models the impact of budget-constrained external agents like election candidates, paid bloggers, or marketing agents on opinion dynamics within a social network. Driven by these practical applications, this book comprehensively explores recent research advancements concerning linear dynamical systems featuring sparse states, control inputs, or both.

Bringing together research from classical control theory and compressed sensing, the book presents the conceptual foundations of sparsity-constrained systems, including the formulation, theory, and algorithms. The approach taken in this work is predominantly mathematical, initially accompanied by a glimpse of its promising practical applications. Our selection of topics was partly influenced by our preferences and prior work while keeping this book an introductory material rather than an exhaustive survey. Nonetheless, the presentation is comprehensive; we present all the necessary background material without assuming any special prior knowledge but basic analysis, linear algebra, and probability theory. We have omitted the proofs of some of the prerequisite results, but provide the corresponding references. Additionally, we have simplified the proofs of the main results, often presenting streamlined versions of those available in existing literature. In this way, our presentation ensures accessibility to students, researchers, or practitioners from the applied mathematics, control system engineering, and signal processing communities.

The outline of the book is as follows. Chapter 1 provides a concise introduction to sparsity in linear dynamical systems, including motivations, applications, and an overview of the entire book. The rest of the book is divided into two parts. The first part comprises Chapters 2 and 3 and delves into the sparse initial state estimation problem, compressive sensing algorithms to solve it, and the analysis of observability under sparsity in the initial state. The second part consists Chapters 4–6 and presents the concept of sparse control, and its effect on the system properties and design algorithms. The book concludes with an appendix covering basic material from matrix analysis, derivation of a Bayesian learning algorithm, and tools from probability theory. Each chapter closes with a "Notes" section, designated for supplemental information that might disrupt the text's flow, such as relevant references, historical insights, additional facts, or open questions.

We hope that readers enjoy reading this book and that their efforts in learning about sparsity-constrained control systems prove to be rewarding.

Delft, The Netherlands Geethu Joseph
Bengaluru, India Chandra R. Murthy

Acknowledgements

We thank all the researchers and their works that have influenced our writing and understanding of the area of sparsity in control linear systems. We have cited these works in the book, but it is quite possible that we have overlooked certain significant works. Our picture of the topic is shaped from various collaborations and discussions. We are particularly indebted to (in alphabetical order) Buddhika Nettasinghe, Chandrasekhar Sriram, Kondapi V. S. Krishna Praveen, Pramod K. Varshney, Rupam Kalyan Chakraborty, and Vikram Krishnamurthy.

We also extend our gratitude to the institutions that supported and provided an excellent working environment during the development of our joint work on this fascinating research area and preparation of this book. We started working together on this area from the Indian Institute of Science (IISc), Bengaluru, India. Later, Geethu Joseph moved to Syracuse University, New York, USA, and then to Delft University of Technology (TU Delft), Netherlands. We acknowledge the financial support from the Intel India Ph.D. fellowship 2015–2019 and IISc-TU Delft Joint Seed Fund 2022–2023.

Contents

About the Authors

Geethu Joseph received the B. Tech. degree in Electronics and Communication Engineering from the National Institute of Technology, Calicut, India, in 2011, and the M. E. degree in Signal Processing and the Ph.D. degree in Electrical Communication Engineering (ECE) from the Indian Institute of Science (IISc), Bangalore, in 2014 and 2019, respectively. She was a postdoctoral fellow with the Department of Electrical Engineering and Computer Science at Syracuse University, NY, USA, from 2019 to 2021. She is currently a tenured assistant professor in the signal processing systems group at the Delft University of Technology, Delft, Netherlands. Dr. Joseph was awarded the 2022 IEEE SPS Best PhD dissertation award and the 2020 SPCOM Best Doctoral Dissertation award. She is also a recipient of the Prof. I. S. N. Murthy Medal in 2014 for being the best M. E. (signal processing) student in the ECE dept., IISc, and the Seshagiri Kaikini Medal for the best Ph.D. thesis of the ECE dept. at IISc for the year 2019–2020. Dr. Joseph holds 35+ peer-reviewed publications in the fields of signal processing, communications, and control theory. She is an associate editor of the IEEE Sensors Journal and an active reviewer for major journals and conferences in signal processing, communications, and control theory. Her research interests include statistical signal processing, network control, and machine learning.

Chandra R. Murthy received his B.Tech. degree in Electrical Engineering from the Indian Institute of Technology (IIT) Madras in 1998, and M.S. and Ph.D. degrees in Electrical and Computer Engineering from Purdue University and the University of California, San Diego, in 2000 and 2006, respectively. In 2007, he joined the Department of Electrical Communication Engineering at the Indian Institute of Science (IISc)Bangalore, India, where he is currently working as a Professor. His research interests are in the areas of sparse control of linear dynamical systems, Bayesian algorithms for sparse signal recovery and their performance analysis, and 5G/6G wireless communications. He has over 75 journals and 100 conference papers to his credit.He is a recipient of the MeitY Young Faculty Fellowship from the Government of India and Prof. Satish Dhawan state award for engineering from the Karnataka State Government. He is a fellow of the IEEE and the Indian National Academy of Engineering (INAE).

Symbols

Vector

$\|\cdot\|$	Euclidean norm of a vector
$\|\cdot\|_0$	Number of nonzero entries of a vector
$\|\cdot\|_1$	ℓ_1−norm of a vector
$\mathrm{supp}(\cdot)$	Support set of a vector
$\mathrm{diag}\{\cdot\}$	Diagonal matrix with the argument along the diagonal

Matrix

$\boldsymbol{A}_{ij}$	(i, j)th entry of the matrix $\boldsymbol{A}$
$\boldsymbol{A}_i$	ith column of the matrix $\boldsymbol{A}$
$\boldsymbol{A}_{\mathcal{S}}$	Submatrix of $\boldsymbol{A}$ with the columns indexed by $\mathcal{S}$
$(\cdot)^\top$	Transpose of a matrix
$(\cdot)^{-1}$	Inverse of a matrix
$\|\cdot\|$	Determinant of a matrix
$(\cdot)^\dagger$	Pseudoinverse of a matrix
$\mathrm{tr}\{\cdot\}$	Trace of a matrix
$\mathrm{rank}\{\cdot\}$	Rank of a matrix
$\|\cdot\|$	Spectral norm of a matrix
$\lambda_{\max}(\cdot)$	Largest eigenvalue of a matrix
$\lambda_{\min}(\cdot)$	Smallest eigenvalue of a matrix
$\mathrm{ColSpace}\{\cdot\}$	Column space of a matrix

Field

$\mathbb{R}$	Field of real numbers
$\mathbb{R}_+$	Field of nonnegative real numbers
$\mathbb{C}$	Field of complex numbers

Set

$\lvert\cdot\rvert$	Cardinality of a set
$(\cdot)^{\complement}$	Complement of a set
$\cup$	Union of two sets
$\cap$	Intersection of two sets

Probability

$\mathbb{P}$	Probability of an event
$\mathbb{E}$	Expectation operator
$\mathcal{N}$	Gaussian distribution

Miscellaneous

$\mathbf{0}$	All zero vector or matrix
$\boldsymbol{I}$	Identity matrix
$\emptyset$	Empty set

Chapter 1
Sparsity in Linear Dynamical Systems

Abstract Linear dynamical systems are widely used models in various control applications, and the control theory for linear systems is a well-studied field. However, many systems face constraints due to limited resources or underlying physical factors, which can be effectively modeled using the concept of sparsity. This chapter introduces the idea of sparsity in linear dynamical systems, highlights its significance across several applications, and provides an overview of the book's content.

Keywords Sparse observability · Sparse controllability · Diffusion sources · Resource-constrained control systems · Compressed sensing · Sparse signal processing

1.1 Sparsity Constraints

Linear dynamical systems are extensively studied and widely embraced mathematical models for describing and analyzing a variety of control systems that evolve over time. The model serves as the core engine across diverse domains such as control systems, signal processing, communications, etc. We represent a discrete linear dynamical system using the state space model,

$$\boldsymbol{x}_k = \boldsymbol{A}\boldsymbol{x}_{k-1} + \boldsymbol{B}\boldsymbol{u}_k, \tag{1.1}$$

for discrete-time indices $k = 1, 2, \ldots$. Here, $\boldsymbol{x}_k$ is a vector that represents the system state at time k. The temporal evolution of the system state is governed by the state transition matrix (also called the state matrix) $\boldsymbol{A}$ and the input matrix $\boldsymbol{B}$. The state $\boldsymbol{x}_k$ at a given time index k depends linearly on the previous state $\boldsymbol{x}_{k-1}$ and is influenced by the input vector $\boldsymbol{u}_k$ applied at time k.

In this book, we explore linear dynamical systems constrained by sparsity, wherein the system state or input vectors or both are assumed to be sparse. A vector is said to be sparse if it contains a large number of zero entries compared to its dimension. Our focus centers on two distinct sparsity models within linear dynamical systems: sparse initial state $\boldsymbol{x}_0$ and sparse control inputs $\boldsymbol{u}_k$'s. Under the sparsity constraints,

G. Joseph and C. R. Murthy, *Sparsity-Constrained Linear Dynamical Systems*,
Springer Tracts in Electrical and Electronics Engineering,
https://doi.org/10.1007/978-981-97-7090-8_1

we delve into two pivotal concepts in control theory: observability and controllability. Informally, observability denotes the ability to estimate the initial state of the system from the observations and known inputs. Controllability, on the other hand, is the ability to steer the system to any desired state using admissible control inputs. This book re-examines observability and controllability through the prism of sparsity. More specifically, we first study the state estimation problem when the initial state of the linear dynamical system is known to be sparse and characterize how the sparsity constraint affects the observability of the system. We next turn to the task of designing sparse control inputs $\boldsymbol{u}_k$'s to drive the system to a desired state and investigate the implications of sparsity constraints on the system's controllability.

Let us notice that it is not immediately clear how to accommodate the knowledge of sparsity in linear dynamical systems due to the temporal dependencies involved. For instance, and as we will see, solving for the *sparsest* initial state is NP-hard in general. However, the emergence of sparsity-aware signal processing algorithms and theory in the past two decades has introduced several fast algorithms with provable guarantees. We study these algorithms and the conditions under which they recover the sparse initial state (as presented in Chapters 2 and 3). Moreover, even if the system is controllable in the classical (non-sparse) sense, with the additional sparsity constraints on the control inputs, the system need not be controllable. It is natural to expect the new notion of controllability to depend on the specific structure of $\boldsymbol{A}$ and $\boldsymbol{B}$ and the dependence is not obvious. As it turns out, the extra condition for controllability under sparsity constraints relies solely on the rank of the system matrix $\boldsymbol{A}$ and is independent of the input matrix $\boldsymbol{B}$ (detailed insights and results presented in Chapter 4). These recent findings have ignited considerable enthusiasm and interest in the fascinating area of sparsity-constrained linear dynamical systems. This book delves deeply into the technical intricacies of these advancements, offering a comprehensive and mathematically rigorous overview.

Before continuing with the discussion on observability and controllability, we comment on the significance of the research topic. We provide several practical examples to motivate the need to study linear dynamical systems under sparsity constraints.

1.2 Sparsity in Initial State

A sparse initial state refers to the initial state of a linear dynamical system where the propagation emerges from a few sources that can spread to the neighboring states in the subsequent time instants. Therefore, the initial state predominantly comprises zero entries, and only a few of its entries are nonzero. Some example systems with sparse initial state are outlined below.

1.2.1 *Diffusion Sources*

Diffusion and propagation processes in large networks starting from a few sources are omnipresent in diverse natural and technological systems. One example is air or water pollution originating from a few sources and spreading across the surrounding area. The initial state is sparse here because the number of sources (or the nonzero entries of the initial state) is limited compared to the polluted area. Other diffusion processes triggered by a few sources are disease or epidemic outbreaks in human society, virus invasion in computer and mobile phone networks, and fake news or rumors spreading on online social networks. In these examples, once the diffusion or propagation emerges, it is critical to locate the diffusion source as quickly as possible to take preventive measures to control further spreading.

1.2.2 *Faulty Events in Power Systems*

Power systems are typically represented using a linear dynamical system where the state is the complex-valued bus voltage vector. In power systems, estimating the network state by monitoring the system is a pertinent research problem. Here, the goal is to identify network events that are typically sparse compared to the number of network buses. Some examples of sparse events are faults (such as shorts to ground, disconnected lines, and line-to-line shorts) and harmonic distortion (sparse in the frequency domain). Thc sparse events can be detected by finding the difference between the consecutive system states. The evolution of difference is also a linear dynamical system but with a sparse (initial) state at the time of the fault event. This sparse initial state estimation is vital to control and maintain voltages within mandated bounds.

The examples above highlight that linear dynamical systems with sparse initial states appear in various systems and elucidate the necessity of determining the sparse state quickly using a small number of observations. Although we do not discuss the scientific details of the above problems, pointers to references treating the topics in much more depth are given in the Notes section at the end of the chapter.

1.3 Sparsity in Control Inputs

Sparse control refers to a control input signal that is sparse in actuator use, i.e., we use a small subset of actuators among the available ones. Specifically, we restrict the control inputs to have very few nonzero entries (or active actuators) compared to their dimension. Sparsity constraints naturally arise in several control systems. We point to a selection of control problems that can be modeled using linear dynamical systems where the input is constrained to be sparse due to cost and energy depletion issues.

1.3.1 Communication-Aware Control

A networked control system refers to a large system where the controlled object(s) and the controller interact over a communication network. For example, consider a drone controlled by a centralized ground controller. The drone sends its sensor data to the ground controller, and based on this information, the controller sends out new control commands to the drone to adjust its position, velocity, and acceleration via the network. Such communication networks are often bandwidth-limited, necessitating sparse control inputs to alleviate communication burdens. The reason for promoting sparsity is that sparse signals admit compact representations and facilitate reduced communication overheads.

1.3.2 Network Opinion Manipulation

Consider a social network where people interact with each other and influence each other's opinions on a product or idea. For example, a social network can be people living in a specific area or social media communities, where a group of people share common interests and experiences. Also, network opinion can refer to a movie rating, inclination towards a political party, or customer rating of a product. The opinion evolution over time can be represented using a linear dynamical system whose state is the network opinion (opinion of all the individuals in the network), and the state matrix models the influence of each individual's opinion on others' opinions. The network information is manipulated by external agents such as paid bloggers, social media influencers, marketing agents, etc. Their influence can be modeled as input to a dynamical system that models opinion evolution. Further, the agents are often constrained by budget (financial or physical), which can be represented using sparse inputs, where sparsity denotes the budget constraints (i.e., the limited reach and influence) of the agent. As an example, consider a company that sends a salesperson to market its products by offering free samples. The number of free samples is limited, and not all the samples reach the target people simultaneously. The influence of such manipulators can modeled using sparse inputs whose nonzero entries correspond to the influence on the individuals who receive the free sample at a given time.

1.3.3 Malicious Data Injection Attacks

In an electric power network, malfunctioning or compromised devices, such as power system stabilizers, generator controllers and exciters, and cyclic loads, can inject forced oscillations (0.1–15 Hz) into the network. The sources triggering these oscillations are typically far fewer than the available sources. Hence, the effect of anomalous sources can be modeled as sparse control inputs to the system. Further, sparse inputs can represent data injection attacks that target a limited number of sensors/actuators in the smart grid. Similarly, sparse inputs can represent malicious attacks on cyber-physical systems.

1.3.4 Efficient Control in Biological Networks

Human metabolism is generally represented by directed networks. Some example biological networks are motivated by the application of control-theoretic ideas in the analysis of biological circuits and biochemical reaction networks. Consider a directed network where the nodes model reactions and/or metabolites. The network can be externally affected by drugs that act only on a few nodes in the network. Here, sparse inputs target to reduce the adverse side effects due to drug administration and it motivates the need to control the system using sparse inputs. As another example, human gait can be modeled using a dynamical system with an input matrix with a large number of columns signifying the different features of the motion. Each motion can be represented as a linear combination of a small number of input matrix columns compared to the total number of columns, leading to a sparse control input.

Overall, the need for sparse control stems from the system's cost constraints or simply from the physics of how the system operates. Therefore, using a limited number of actuators is desirable without significantly compromising the control performance, for example, in terms of the time or energy required to reach a certain desired state.

1.4 Control Theory with Sparsity Constraints

The classical control theory of observability and controllability of a linear dynamical system without any (sparsity) constraints is a mature and well-established research area. It is also known that observability and controllability are sister notions and dual problems, and their analysis is quite similar. However, the sparsity constraints on initial state and/or control inputs alters the duality relationships. Specifically, the dual of observability of the sparse initial state is the reachability of the system to sparse final states. Similarly, the dual of controllability using sparse inputs is observability using a small subset of observations. Therefore, observability and controllability under sparsity constraints on initial state and control inputs, respectively, are two different problems and demand new mathematical analysis.

Furthermore, the two sparsity constraints lead to two types of restrictions on the system. When the initial state is known to be sparse, we effectively have additional information about the unknown state of the system. Therefore, if the underlying linear dynamical system is observable, the sparse initial state becomes observable as well. Hence, classical observability is a sufficient condition for the observability of a sparse initial state. As we will show later, the systems deemed unobservable using classical control theory can become observable when the additional sparsity structure of the unknown initial state is leveraged in the estimation method. Moreover, when the inputs are confined to be sparse, the set of admissible inputs becomes restricted compared to the classical setting. Consequently, classical controllability is a prerequisite for achieving controllability using sparse inputs. As a result, the systems that are controllable using classical control theory can be uncontrollable under additional sparsity constraints. This simple observation further highlights the distinct nature of the two problems and the need for separate analyses.

We underscore that the primary challenge in sparse initial state estimation or sparse control design lies in identifying the small subsets of nonzero entries in the unknowns. If the subset is known, the estimation or design problem reduces to the standard initial state estimation or control input design problem. Identifying this subset is a combinatorial problem, and the sparsity constraint is nonconvex. Consequently, the analysis and design of algorithms and guarantees adhering to these sparsity constraints are far from trivial. This book is devoted to the fundamental limits, mathematical tools, and algorithms for sparsity-aware estimation in linear dynamical systems.

1.5 Overview of the Book

This book is organized into two parts (excluding this introductory chapter). The problem formulation and associated analysis are readily accessible to signal processing, control/systems theory, and applied mathematics communities.

The first part has two chapters (Chapters 2 and 3) and deals with sparse state estimation.

Chapter 2 defines the notion of vector sparsity and formulates the sparse initial state estimation problem. This chapter answers the following question:

How do we estimate a sparse initial state of a linear dynamical system through a computationally tractable reconstruction algorithm?

First, we identify that the sparse initial state estimation problem is similar to the compressed sensing problem. A few popular algorithms used in compressive sensing that can be used to estimate the sparse initial state are presented. The algorithms are divided into four categories: optimization methods, greedy methods, thresholding-based methods, and Bayesian methods. Their rigorous analyses are postponed to the next chapter.

Chapter 3 quantifies the number of measurements required to successfully recover a sparse initial state of a given linear dynamical system. Specifically, we ask

What are sufficient conditions for the recoverability of sparse initial states via the compressed sensing algorithms?

This chapter starts with classical results on the initial state recoverability, also called observability. Then, we provide a brief introduction to the mathematical concept of restricted isometry property used for our sparse observability analysis. We then investigate the conditions on a stochastic linear dynamical system that ensure the recoverability of a sparse initial state and characterize the corresponding observability index. The results establish that a sparse initial state can be estimated with far fewer observations (that scales logarithmically with the state dimension) than required by an unstructured (non-sparse) initial state.

The second part of the book comprises three chapters (Chapters 4 to 6) and provides a detailed study of sparse control.

Chapter 4 formally introduces the notion of sparse control and defines the notion of controllability under the sparsity constraint. The central questions of the chapter are as follows:

What are necessary and sufficient conditions for ensuring controllability under sparse inputs with possibly different supports? Can we devise a computationally simple test for sparse controllability? If a system is controllable using sparse inputs, how many control input vectors are needed to drive the system from a given initial state to an arbitrary final state?

We show that, for any linear dynamical system controllable under the sparsity constraint, a sparse actuator schedule independent of the system state exists, which can drive the system to any desired state. Further, we derive simple algebraic conditions, which are both necessary and sufficient for the sparse controllability of the system. We show that the system is sparse controllable if and only if it is controllable *and* the sparsity level exceeds the nullity of the state matrix. Unlike the more traditional Kalman-type rank tests, the derived conditions can be verified in polynomial time complexity. Finally, we characterize the time-to-control or the minimum number of input vectors required to ensure sparse controllability and show that the state dimension bounds it. These results form a theoretical basis for designing sparse control inputs, which we discuss in the next chapter.

Chapter 5 addresses the design of sparse control with time-varying support for a given linear dynamical system. This chapter seeks an answer to the following question:

Given a controllable linear dynamical system, how do we design sparse inputs that take the system from a given initial state to a desired final state?

We formulate the sparse control input design in two ways. In the first approach, we formulate it as a sparse recovery problem and use the compressed sensing algorithms to solve it. This approach does not necessarily assume that the system is controllable, but the corresponding sparse actuator schedule depends on the initial and final states. In the second approach, we assume that the system is controllable. We design a global sparse actuator schedule that applies to any pair of initial and final states and then derive the control inputs based on the designed schedule.

Chapter 6 extends the idea of sparse control to other related control theory notions. The focal question addressed in the chapter is as follows:

> *How are the stabilizability, output controllability, and nonnegative controllability of a linear dynamical system affected by the sparsity constraints on the input?*

We show three key results in this chapter. The first result, perhaps surprisingly, shows that sparsity constraints do not have any effect on the stabilizability. All stabilizable systems are sparse stabilizable. The second result is on the algebraic characterization of sparse output controllability. We derive bounds on minimal sparsity levels that ensure output controllability. Finally, we show that sparse controllable and nonnegative controllable systems are sparse nonnegative controllable. We also briefly discuss three sets of design algorithms: the first estimates sparse control inputs for system stabilization, the second estimates sparse control inputs to achieve a desired output, and the third estimates nonnegative sparse control inputs to reach a desired state.

Each chapter ends with a chapter summary and a section titled "Notes", where we provide relevant references, additional comments, and list some open questions.

Notation

Notation is usually introduced when it is used for the first time. The collection of symbols used in the text can be found at the beginning of the book.

Out-of-the-Scope Topics

This book, while not exhaustive, presents the key recent research on linear dynamical systems and sparsity, connecting control theory and compressed sensing (sparse signal recovery). Other important problems related to sparsity and control have been studied in the literature but are outside the scope of this book.

Example topics related to sparse observability are the estimation of sparse states of linear dynamical systems using filtering methods and the joint estimation of states and sparse inputs Sefati et al. (2015); Vaswani (2008); Poe et al. (2023); Joseph and Murthy (2017); Joseph and Varshney (2022); Chakraborty et al. (2024); Prasad et al. (2014).

Some left-out topics related to sparse controllability are sparsity in feedback to the controller (minimizing the feedback bandwidth) Babazadeh and Nobakhti (2016); Arastoo et al. (2016); Lin et al. (2013); Polyak et al. (2013, 2014); Lin et al. (2012);

Wang et al. (2014); Bahavarnia et al. (2017), sparsity in time (minimizing the duration over which a nonzero control is applied) Nagahara (2023); Chatterjee et al. (2016); Ikeda and Kashima (2018b); Nagahara (2020); Ito et al. (2021); Olshevsky (2020); Ikeda and Kashima (2018a), and minimal input selection for (strong) structural controllability Chapman and Mesbahi (2013); Abbas et al. (2023); Trefois and Delvenne (2015).

Interested readers are referred to the above works and references therein for an extensive treatment of the topics.

Notes

The linear dynamical systems are widely used in diverse areas such as control systems Zhou et al. (1996), signal processing Anderson and Moore (2005), communications Prasad et al. (2014), economics Brockwell et al. (2002), mechanical and civil engineering Pope III et al. (2002); Shao et al. (2006), health care Neumann et al. (2009); Hvistendahl et al. (2013), etc. The study of a linear dynamical system with sparsity constraints has roots dating back to 1972 Athans (1972). However, sparsity leads to discrete constraints, making the analysis cumbersome.

With the emergence of compressed sensing theory in the last two decades, new computationally efficient algorithms and sophisticated analytical tools to handle sparsity have been introduced in the literature. The field of compressive sensing took its initial steps through seminal papers in the early 2000s, marking a significant juncture by combining the concepts of ℓ_1-norm minimization with randomness in the measurement matrices Candès et al. (2006); Donoho (2006). For comprehensive overviews on compressive sensing, see Baraniuk (2007); Wakin et al. (2008); Kutyniok (2013); Fornasier and Rauhut (2015); Foucart and Rauhut (2013). Drawing inspiration from the area of compressed sensing, some recent works have addressed the problems of sparse initial state estimation and sparse control input design, as well as the other related problems that we cover in this book.

The sparse initial state estimation in linear dynamical systems is motivated by disease or epidemic spreading in the human society Neumann et al. (2009); Hvistendahl et al. (2013), pollution Pope III et al. (2002); Rumchev and Caccetta (2007), virus spreading in computer and mobile phone networks Gan et al. (2012); Lloyd and May (2001); Wang et al. (2009), and information propagation in online platforms Cha et al. (2009); DeGroot (1974); Das et al. (2014); Joseph et al. (2021). The sparse state estimation problem in power systems is presented in Rozenberg and Levron (2016); Rozenberg et al. (2018). The sparse observability analysis can be found in Sanandaji et al. (2014); Joseph and Murthy (2018).

Further, sparse control for the network control systems is inspired by communication constraints presented in Heemels et al. (2010); Tatikonda and Mitter (2004); Liu et al. (2020); Nagahara and Quevedo (2011). Network opinion manipulation using sparse control was introduced in Joseph et al. (2021). The sparse sources of forced oscillations in electric power networks are discussed in Anguluri et al. (2023, 2022); Siami et al. (2020). Sparse control for biological networks is motivated by the application of control-theoretic ideas in the analysis of biological circuits Marucci et al. (2009), biochemical reaction networks Liu et al. (2013), systems biology Rajapakse et al. (2012), and sparse representations of human gait Bissacco et al. (2001); Raptis et al. (2010). Sparse data injection attacks in smart grids and cyber-physical systems are discussed in Cárdenas et al. (2008); Hao et al. (2015); Sun and Li (2022); Chen et al. (2019); Ma and Shi (2022); Tsang et al. (2020). Linear dynamical systems with sparse control inputs also model several other practical applications like seismic signal processing Chapman and Barrodale (1983); Carpentier et al. (2018), which we have not covered in this book. Sparse control-related problems are analyzed in Joseph and Murthy (2020); Siami and Jadbabaie (2018).

References

Abbas W, Shabbir M, Yazıcıoğlu Y, Koutsoukos X (2023) On zero forcing sets and network controllability–computation and edge augmentation. IEEE Trans Control Netw Syst

Anderson B, Moore J (2005) Optimal filtering. Courier Dover

Anguluri R, Kosut O, Sankar L (2023) Localization and estimation of unknown forced inputs: a group LASSO approach. IEEE Trans Control Netw Syst

Anguluri R, Taghipourbazargani N, Kosut O, Sankar L (2022) A complex-LASSO approach for localizing forced oscillations in power systems. In: Proceedings of IEEE power & energy society general meeting, pp 01–05

Arastoo R, GhaedSharaf Y, Kothare MV, Motee N (2016) Optimal state feedback controllers with strict row sparsity constraints. In: Proceedings of American control conference, pp 1948–1953

Athans M (1972) On the determination of optimal costly measurement strategies for linear stochastic systems. Automatica 8(4):397–412

Babazadeh M, Nobakhti A (2016) Sparsity promotion in state feedback controller design. IEEE Trans Autom Control 62(8):4066–4072

Bahavarnia M, Somarakis C, Motee N (2017) State feedback controller sparsification via a notion of non-fragility. In: Proceedings of IEEE conference on decision and control, pp 4205–4210

Baraniuk RG (2007) Compressive sensing [lecture notes]. IEEE Signal Process Mag 24(4):118–121

Bissacco A, Chiuso A, Ma Y, Soatto S (2001) Recognition of human gaits. In: Proceedings of IEEE computer society conference on computer vision and pattern recognition, vol 2, pp II–II

Brockwell PJ, Davis RA, Calder MV (2002) Introduction to time series and forecasting, vol 2. Springer

Candès EJ, Romberg J, Tao T (2006) Robust uncertainty principles: exact signal reconstruction from highly incomplete frequency information. IEEE Trans Inf Theory 52(2):489–509

Cárdenas AA, Amin S, Sastry S (2008) Research challenges for the security of control systems. Hot Top Secur 5(15):1158

Carpentier S, Abidin H, Steeghs P, Veldkamp H (2018) Identifying hidden risk elements for co2 storage from reprocessed seismic data. In: Proceedings of CO2 geological storage workshop, European association of geoscientists & engineers, pp 1–5

Chakraborty RK, Joseph G, Murthy CR (2024) Joint state and input estimation for linear dynamical systems with sparse control. In: Proceedings of international conference on acoustics, speech, and signal processing

Cha M, Mislove A, Gummadi KP (2009) A measurement-driven analysis of information propagation in the flickr social network. In: Proceedings of the international conference on world wide web, pp 721–730

Chapman N, Barrodale I (1983) Deconvolution of marine seismic data using the l 1 norm. Geophys J Int 72(1):93–100

Chapman A, Mesbahi M (2013) On strong structural controllability of networked systems: a constrained matching approach. In: Proceedings of American control conference, pp 6126–6131
Chatterjee D, Nagahara M, Quevedo DE, Rao KM (2016) Characterization of maximum hands-off control. Syst & Control Lett 94:31–36
Chen R, Li X, Zhong H, Fei M (2019) A novel online detection method of data injection attack against dynamic state estimation in smart grid. Neurocomputing 344:73–81
Das A, Gollapudi S, Munagala K (2014) Modeling opinion dynamics in social networks. In: Proceedings of the ACM international conference on web search and data mining, pp 403–412
DeGroot MH (1974) Reaching a consensus. J Am Stat Assoc 69(345):118–121
Donoho DL (2006) Compressed sensing. IEEE Trans Inf Theory 52(4):1289–1306
Fornasier M, Rauhut H (2015) Compressive sensing. In: Handbook of mathematical methods in imaging, vol 1, pp 187–229
Foucart S, Rauhut H (2013) An invitation to compressive sensing. Springer
Gan C, Yang X, Liu W, Zhu Q, Zhang X, et al (2012) Propagation of computer virus under human intervention: a dynamical model. Discret Dyn Nat Soc 2012
Hao J, Piechocki RJ, Kaleshi D, Chin WH, Fan Z (2015) Sparse malicious false data injection attacks and defense mechanisms in smart grids. IEEE Trans Ind Inform 11(5):1–12
Heemels WMH, Teel AR, Van de Wouw N, Nešić D (2010) Networked control systems with communication constraints: tradeoffs between transmission intervals, delays and performance. IEEE Trans Autom Control 55(8):1781–1796
Hvistendahl M, Normile D, Cohen J (2013) Despite large research effort, H7N9 continues to baffle. Science 340(6131):414–415
Ikeda T, Kashima K (2018) Sparsity-constrained controllability maximization with application to time-varying control node selection. IEEE Control Syst Lett 2(3):321–326
Ikeda T, Kashima K (2018) On sparse optimal control for general linear systems. IEEE Trans Autom Control 64(5):2077–2083
Ito K, Ikeda T, Kashima K (2021) Sparse optimal stochastic control. Automatica 125:109438
Joseph G, Murthy CR (2017) A noniterative online Bayesian algorithm for the recovery of temporally correlated sparse vectors. IEEE Trans Signal Process 65(20):5510–5525
Joseph G, Murthy CR (2018) On the observability of a linear system with a sparse initial state. IEEE Signal Process Lett 25(7):994–998
Joseph G, Murthy CR (2020) Controllability of linear dynamical systems under input sparsity constraints. IEEE Trans Autom Control 66(2):924–931
Joseph G, Nettasinghe B, Krishnamurthy V, Varshney PK (2021) Controllability of network opinion in erdös-rènyi graphs using sparse control inputs. SIAM J Control Optim 59(3):2321–2345
Joseph G, Varshney PK (2022) State estimation of linear systems with sparse inputs and Markov-modulated missing outputs. In: Proceedings of European signal processing conference, pp 837–841
Kutyniok G (2013) Theory and applications of compressed sensing. GAMM-Mitteilungen 36(1):79–101
Lin F, Fardad M, Jovanović MR (2013) Design of optimal sparse feedback gains via the alternating direction method of multipliers. IEEE Trans Autom Control 58(9):2426–2431
Lin F, Fardad M, Jovanović MR (2012) Sparse feedback synthesis via the alternating direction method of multipliers. In: Proceedings of American control conference, pp 4765–4770
Liu YY, Slotine JJ, Barabási AL (2013) Observability of complex systems. Proc Natl Acad Sci 110(7):2460–2465
Liu K, Fridman E, Xia Y (2020) Networked control under communication constraints. Springer
Lloyd AL, May RM (2001) How viruses spread among computers and people. Science 292(5520):1316–1317
Ma R, Shi P (2022) Secure state estimation for cyber-physical systems under sparse data injection attacks: a switched counteraction approach. Int J Control 95(4):963–974

Marucci L, Barton DA, Cantone I, Ricci MA, Cosma MP, Santini S, di Bernardo D, di Bernardo M (2009) How to turn a genetic circuit into a synthetic tunable oscillator, or a bistable switch. PloS one 4(12):e8083

Nagahara M (2020) Sparsity methods for systems and control. Now Publishers

Nagahara M (2023) Sparse control for continuous-time systems. Int J Robust Nonlinear Control 33(1):6–22

Nagahara M, Quevedo DE (2011) Sparse representations for packetized predictive networked control. IFAC Proc Vol 44(1):84–89

Neumann G, Noda T, Kawaoka Y (2009) Emergence and pandemic potential of swine-origin H1N1 influenza virus. Nature 459(7249):931–939

Olshevsky A (2020) On a relaxation of time-varying actuator placement. IEEE Control Syst Lett 4(3):656–661

Poe K, Mallada E, Vidal R (2023) Necessary and sufficient conditions for simultaneous state and input recovery of linear systems with sparse inputs by ℓ_1-minimization. arXiv:2304.05526

Polyak BT, Khlebnikov MV, Shcherbakov PS (2014) Sparse feedback in linear control systems. Autom Remote Control 75:2099–2111

Polyak B, Khlebnikov M, Shcherbakov P (2013) An LMI approach to structured sparse feedback design in linear control systems. In: Proceedings of European control conference, pp 833–838

Pope CA III, Burnett RT, Thun MJ, Calle EE, Krewski D, Ito K, Thurston GD (2002) Lung cancer, cardiopulmonary mortality, and long-term exposure to fine particulate air pollution. J Am Med Assoc 287(9):1132–1141

Prasad R, Murthy C, Rao B (2014) Joint approximately sparse channel estimation and data detection in OFDM systems using sparse Bayesian learning. IEEE Trans Signal Process 62(14):3591–3603

Rajapakse I, Groudine M, Mesbahi M (2012) What can systems theory of networks offer to biology? PLoS Comput Biol 8(6):e1002543

Raptis M, Wnuk K, Soatto S (2010) Spike train driven dynamical models for human actions. In: Proceedings of IEEE computer society conference on computer vision and pattern recognition, pp 2077–2084

Rozenberg I, Beck Y, Eldar YC, Levron Y (2018) Sparse estimation of faults by compressed sensing with structural constraints. IEEE Trans Power Syst 33(6):5935–5944

Rozenberg I, Levron Y (2016) Observability challenges in sparse estimation of fault events. In: Proceedings of IEEE international energy conference, pp 1–6

Rumchev V, Caccetta L (2007) A dynamic model of mobile source air pollution and its properties. WIT Trans Biomed Health 11

Sanandaji BM, Wakin MB, Vincent TL (2014) Observability with random observations. IEEE Trans Autom Control 59(11):3002–3007

Sefati S, Cowan NJ, Vidal R (2015) Linear systems with sparse inputs: observability and input recovery. In: Proceedings of American control conference, pp 5251–5257

Shao M, Tang X, Zhang Y, Li W (2006) City clusters in China: air and surface water pollution. Front Ecol Environ 4(7):353–361

Siami M, Olshevsky A, Jadbabaie A (2020) Deterministic and randomized actuator scheduling with guaranteed performance bounds. IEEE Trans Autom Control 66(4):1686–1701

Siami M, Jadbabaie A (2018) Deterministic polynomial-time actuator scheduling with guaranteed performance. In: Proceedings of European control conference, pp 113–118

Sun K, Li Z (2022) Sparse data injection attacks on smart grid: an information-theoretic approach. IEEE Sens J 22(14):14553–14562

Tatikonda S, Mitter S (2004) Control under communication constraints. IEEE Trans Autom Control 49(7):1056–1068

Trefois M, Delvenne JC (2015) Zero forcing number, constrained matchings and strong structural controllability. Linear Algebr Appl 484:199–218

Tsang KFE, Huang M, Johansson KH, Shi L (2020) Sparse linear injection attack on multi-agent consensus control systems. IEEE Control Syst Lett 5(2):665–670

Vaswani N (2008) Kalman filtered compressed sensing. In: Proceedings of IEEE international conference on image processing, pp 893–896

Wakin MB et al (2008) An introduction to compressive sampling. IEEE Signal Process Mag 25(2):21–30

Wang P, González MC, Hidalgo CA, Barabási AL (2009) Understanding the spreading patterns of mobile phone viruses. Science 324(5930):1071–1076

Wang Y, Lopez J, Sznaier M (2014) Sparse static output feedback controller design via convex optimization. In: Proceedings of IEEE conference on decision and control, pp 376–381

Zhou K, Doyle JC, Glover K (1996) Robust and optimal control, vol 40. Prentice Hall, New Jersey

Chapter 2
Sparse Initial State Estimation Algorithms

Abstract In this chapter, we estimate the sparse initial state of a linear dynamical system from its observations when the inputs are known. We formulate this as a sparse vector recovery problem, a topic well-studied in the signal processing field of compressed sensing. We present several sparsity-aware algorithms from the compressed sensing literature to address this problem, including popular methods such as convex relaxation-based basis pursuit, greedy algorithms like orthogonal matching pursuit and compressive sampling matching pursuit, as well as thresholding techniques, such as iterative hard thresholding, hard thresholding pursuit, and sparse Bayesian learning.

Keywords Sparse vector recovery · Basis pursuit · Orthogonal matching pursuit · Compressive sampling matching pursuit · Iterative hard thresholding · Hard thresholding pursuit · Sparse Bayesian learning

2.1 Sparsity and Sparse Initial State

We first define the notion of sparsity and support set.

Definition 2.1 The support of a vector $\boldsymbol{x} \in \mathbb{R}^N$, denoted as $\text{supp}(\boldsymbol{x})$ is the index set of its nonzero values,

$$\text{supp}(\boldsymbol{x}) = \{j : \boldsymbol{x}_j \neq 0\}. \tag{2.1}$$

A vector is said to be s-sparse if it has at most s nonzero entries,

$$\|\boldsymbol{x}\|_0 = |\text{supp}(\boldsymbol{x})| \leq s, \tag{2.2}$$

where $\|\cdot\|_0$ is the ℓ_0 norm that counts the number of nonzero entries in a vector and $|\cdot|$ denotes the cardinality of a set.

To define sparsity in the initial state, we recall the state space model of the linear dynamical system ($\boldsymbol{A} \in \mathbb{R}^{n\times n}$, $\boldsymbol{B} \in \mathbb{R}^{n\times m}$, $\boldsymbol{C}_{(k)} \in \mathbb{R}^{p\times m}$, $k = 0, 1, 2, \ldots$), with state $\boldsymbol{x}_k \in \mathbb{R}^n$, input $\boldsymbol{u}_k \in \mathbb{R}^m$, and output $\boldsymbol{y}_k \in \mathbb{R}^p$,

G. Joseph and C. R. Murthy, *Sparsity-Constrained Linear Dynamical Systems*, Springer Tracts in Electrical and Electronics Engineering,
https://doi.org/10.1007/978-981-97-7090-8_2

$$\boldsymbol{x}_k = \boldsymbol{A}\boldsymbol{x}_{k-1} + \boldsymbol{B}\boldsymbol{u}_k, \quad \text{and} \quad \boldsymbol{y}_k = \boldsymbol{C}_{(k)}\boldsymbol{x}_k, \tag{2.3}$$

at discrete time index k. Here, $\boldsymbol{C}_{(k)}$ is the observation matrix at time k. Here, we consider time-varying matrices; note that a time-invariant system is a special case of the above time-varying system. We focus on the case where $p \ll n$, implying that the outputs are (very) low-dimensional linear projections (compressive measurements) of the corresponding state vectors. Also, we assume that the initial state $\boldsymbol{x}_0$ is sparse, i.e., $\|\boldsymbol{x}_0\|_0 \ll n$.

Suppose we apply control inputs to a linear dynamical system and observe the output of the system up to a finite time K. We are interested in recovering the sparse initial state $\boldsymbol{x}_0$ from the low-dimensional observations $\boldsymbol{y}_k$, with $k = 1, 2, \ldots, K$. If any initial state can be determined by examining its outputs for some $K < \infty$, then the system is said to be observable. Also, the observability of the initial state ensures the observability of the system state at any time k when the inputs are known. So, the notion of observability characterizes the visibility into the system states and is important to analyze and optimize linear dynamical systems. The mathematical definition of observability and its associated analysis is presented in Chapter 3. In this chapter, we formulate the sparse state recovery problem and present algorithms that estimate the initial state by exploiting sparsity.

2.2 Sparse Initial State Estimation

To recover the sparse initial state, we rewrite the output at time k in (2.3) as

$$\boldsymbol{y}_k = \boldsymbol{C}_{(k)}\boldsymbol{x}_k = \boldsymbol{C}_{(k)}\left[\boldsymbol{A}\boldsymbol{x}_{k-1} + \boldsymbol{B}\boldsymbol{u}_k\right] \tag{2.4}$$

$$= \boldsymbol{C}_{(k)}\left[\boldsymbol{A}^k\boldsymbol{x}_0 + \sum_{j=1}^{k} \boldsymbol{A}^{k-j}\boldsymbol{B}\boldsymbol{u}_j\right]. \tag{2.5}$$

We can group the known quantities to one side to get the system of equations,

$$\boldsymbol{y}_k - \boldsymbol{C}_{(k)}\sum_{j=1}^{k} \boldsymbol{A}^{k-j}\boldsymbol{B}\boldsymbol{u}_j = \boldsymbol{C}_{(k)}\boldsymbol{A}^k\boldsymbol{x}_0. \tag{2.6}$$

Stacking the K outputs at time instants $k = 0, 1, \ldots, K-1$, we arrive at the following equivalent system of equations:

$$y^{(K)} = \begin{bmatrix} y_0 \\ y_1 \\ \vdots \\ y_{K-1} \end{bmatrix} - \begin{bmatrix} \mathbf{0} & & & \\ C_{(1)}B & & & \\ C_{(2)}AB & C_{(2)}B & & \\ \vdots & \vdots & \ddots & \\ C_{(K-1)}A^{K-2}B & C_{(K-1)}A^{K-3}B & \dots & C_{(K-1)}B \end{bmatrix} \begin{bmatrix} u_1 \\ u_2 \\ \vdots \\ u_{K-1} \end{bmatrix} \tag{2.7}$$

$$= \begin{bmatrix} C_{(0)} \\ C_{(1)}A \\ \vdots \\ C_{(K-1)}A^{K-1} \end{bmatrix} x_0 \tag{2.8}$$

$$= O^{(K)} x_0, \tag{2.9}$$

where $O^{(K)} \in \mathbb{R}^{pK \times n}$ is called the *observability matrix* obtained using K output vectors. Our goal is to solve the system of equations, $y^{(K)} = O^{(K)} x_0$, to estimate the initial state x_0.

If x_0 is arbitrary, we need the number of equations to be greater than or equal to the number of unknowns, $pK \geq n$, to recover x_0. If the system has a unique solution, we can use Gaussian elimination or similar approaches to solve for x_0. However, if the initial state is known to be sparse, i.e., $\|x_0\|_0 \ll n$, we can leverage this knowledge to correctly determine x_0 even when the system of linear equations is underdetermined, $pK < n$. Signal processing techniques for efficiently reconstructing sparse vectors by finding solutions to underdetermined linear systems is one of the core topics of interest in compressive sensing (also called compressed sensing, compressive sampling, sparse sampling, or sparse signal recovery). We discuss the basics of compressed sensing and the related algorithms in the following sections.

2.3 Compressive Sensing Algorithms

The compressive sensing problem refers to solving for an s-sparse vector $x_0 \in \mathbb{R}^n$ satisfying the underdetermined system of linear equations,

$$y^{(K)} = O^{(K)} x_0, \tag{2.10}$$

where $y^{(K)} \in \mathbb{R}^{pK \times 1}$ and the measurement matrix $O^{(K)} \in \mathbb{R}^{pK \times n}$, with $pK < n$. If x_0 is the unique sparsest solution to (2.10), we can recover it by solving the following optimization problem,

$$\underset{z \in \mathbb{R}^n}{\arg\min} \; \|z\|_0 \quad \text{s. t.} \quad O^{(K)} z = y^{(K)}. \tag{2.11}$$

However, since ℓ_0-norm is nonconvex, it is not easy to directly solve (2.11). A straightforward approach is to start with $\|z\|_0 = 1$, and consider finding a solution to $\boldsymbol{O}_{\mathcal{S}}^{(K)} z_{\mathcal{S}} = \boldsymbol{y}^{(K)}$ for all possible support sets $\mathcal{S}$ with $|\mathcal{S}| = \|z\|_0$. If none of the systems of equations has a solution, we increase $\|z\|_0$ by 1 and continue until we find a solution. This combinatorial search finds the exact solution to the problem. However, due to the enormous number of subsets $\mathcal{S}$ involved, which is of exponential complexity in the number of nonzero entries, it is impractical to solve (2.11) using this direct approach even for moderate-sized problems.

In fact, the general ℓ_0-norm minimization problem is difficult to solve using any practical approaches. Fortunately, the compressed sensing literature offers several tractable algorithms that can provably determine $\boldsymbol{x}_0$ from (2.11) when certain conditions, for example, on the observability matrix, are satisfied. However, we reiterate these algorithms are not guaranteed to succeed for all possible choices of linear dynamical systems and initial states.

In the subsequent sections, we present a representative selection of popular sparse recovery algorithms used in compressive sensing. We present four types of algorithms: convex optimization, greedy, thresholding, and Bayesian methods. Generally, the algorithms outlined in this chapter tend to perform effectively in practical scenarios. The selection of the best algorithm among them depends on specific system parameters and requirements like the desired accuracy level, the number of measurements, the sparsity level, computational complexity, etc.

2.3.1 Convex Optimization Method

We note that the optimization problem in (2.11),

$$\arg\min_{z \in \mathbb{R}^n} \|z\|_0 \quad \text{s. t.} \quad \boldsymbol{y}^{(K)} = \boldsymbol{O}^{(K)} z, \tag{2.12}$$

is nonconvex due to the ℓ_0-norm in the objective function. Nonetheless, the ℓ_q norm $\|z\|_q = \left(\sum_i |z_i|^q\right)^{1/q}$ approaches $\|z\|_0$, when $q > 0$ is close to 0. Consequently, we can approximate (2.12) as

$$\arg\min_{z \in \mathbb{R}^n} \|z\|_q \quad \text{s. t.} \quad \boldsymbol{y}^{(K)} = \boldsymbol{O}^{(K)} z, \tag{2.13}$$

for some $q > 0$. For $q < 1$, the problem in (2.13) is nonconvex and hard to solve. However, when $q \geq 1$, the problem is convex and can be solved using any standard convex optimization solver. So we choose the smallest value, $q = 1$, for which the objective function is convex, and the resulting optimization problem is an approximation to solve (2.12),

$$\arg\min_{z \in \mathbb{R}^n} \|z\|_1 \quad \text{s. t.} \quad \boldsymbol{y}^{(K)} = \boldsymbol{O}^{(K)} z. \tag{2.14}$$

The above principle is usually called ℓ_1-minimization or basis pursuit, and the recipe is summarized in Algorithm 2.1.

Algorithm 2.1 Sparse initial state estimation via basis pursuit

Input: Outputs $\{y_k \in \mathbb{R}^p\}_{k=0}^{K-1}$, control inputs $\{u_k \in \mathbb{R}^m\}_{k=1}^{K-1}$ linear dynamical system ($A \in \mathbb{R}^{n\times n}$, $B \in \mathbb{R}^{n\times m}$, $C_{(k)} \in \mathbb{R}^{p\times m}$, $k = 0, 1, \ldots, K-1$)
1: Compute $y^{(K)}$ and $O^{(K)}$ using (2.7) and (2.9)
2: Determine x_0 by solving the convex optimization problem (2.14)
Output: Initial state x_0.

2.3.2 Greedy Algorithms

We next present two greedy algorithms to solve (2.12), namely orthogonal matching pursuit and compressive sampling matching pursuit. Both algorithms are iterative in nature and update the support set $\mathcal{S}$ of z in every iteration using greedy heuristics.

The orthogonal matching pursuit algorithm starts with an empty support set $\mathcal{S}$ and $z = \mathbf{0}$. In every iteration, it computes the residue, $y^{(K)} - O^{(K)}z$, and matches it with the columns of $O^{(K)}$ that are not indexed by $\mathcal{S}$. The column with the highest inner product (in absolute value) is added to the target support $\mathcal{S}$. Using the updated support, the corresponding estimate of z is computed. These steps are continued until $y^{(K)} = O^{(K)}z$ is satisfied. The resulting algorithm is summarized in Algorithm 2.2, where $(\cdot)^\dagger$ denotes the pseudoinverse of a matrix.

Algorithm 2.2 Sparse initial state estimation via orthogonal matching pursuit

Input: Outputs $\{y_k \in \mathbb{R}^p\}_{k=0}^{K-1}$, control inputs $\{u_k \in \mathbb{R}^m\}_{k=1}^{K-1}$ linear dynamical system ($A \in \mathbb{R}^{n\times n}$, $B \in \mathbb{R}^{n\times m}$, $C_{(k)} \in \mathbb{R}^{p\times m}$, $k = 0, 1, \ldots, K-1$)
Initialization: $\mathcal{S} = \emptyset$, $z = \mathbf{0}$
1: Compute $y^{(K)}$ and $O^{(K)}$ using (2.7) and (2.9)
2: **repeat**
3: Compute $i^* = \arg\max_{i \notin \mathcal{S}} \frac{\left| O_i^{(K)\mathsf{T}} (y^{(K)} - O^{(K)} z) \right|}{\left\| O_i^{(K)} \right\|}$
4: Update support $\mathcal{S} \leftarrow \mathcal{S} \cup \{i^*\}$
5: Determine $z_{\mathcal{S}} \leftarrow O_{\mathcal{S}}^{(K)\dagger} y^{(K)}$ and $z_{\mathcal{S}^{\complement}} \leftarrow \mathbf{0}$
6: **until** $y^{(K)} = O^{(K)}z$
Output: Initial state $x_0 = z$

The orthogonal matching pursuit algorithm generally works well. However, a drawback of the algorithm is that once an incorrect index is added to the target support $\mathcal{S}$, it cannot be removed. The second greedy algorithm, called the compressive sampling matching pursuit algorithm, addresses this issue. Every iteration of compressive sampling matching pursuit first adds the indices of top $2s$ columns of $\boldsymbol{O}^{(K)}$ that match with the residue. Then, the corresponding estimate of z is computed using the extended target support. Finally, it keeps the estimate's s largest absolute entries and sets the remaining entries to zero. In effect, every iteration of the algorithm returns an s-sparse vector. The iterations are stopped when (2.10) holds. The resulting algorithm is summarized in Algorithm 2.3.

Algorithm 2.3 Sparse initial state estimation via compressive sampling matching pursuit

---

Input: Outputs $\{\boldsymbol{y}_k \in \mathbb{R}^p\}_{k=0}^{K-1}$, control inputs $\{\boldsymbol{u}_k \in \mathbb{R}^m\}_{k=1}^{K-1}$ linear dynamical system ($\boldsymbol{A} \in \mathbb{R}^{n\times n}$, $\boldsymbol{B} \in \mathbb{R}^{n\times m}$, $\boldsymbol{C}_{(k)} \in \mathbb{R}^{p\times m}$, $k = 0, 1, \ldots, K-1$), sparsity level s

Initialization: $\mathcal{S} = \emptyset$, $z = \boldsymbol{0}$

1: Compute $\boldsymbol{y}^{(K)}$ and $\boldsymbol{O}^{(K)}$ using (2.7) and (2.9)

2: **repeat**

3: Compute $\mathcal{I}^* = \underset{\substack{\mathcal{I}\subseteq\{1,2,\ldots,n\}\\|\mathcal{I}|\leq 2s}}{\arg\max} \sum_{i\in\mathcal{I}} \frac{\left|\boldsymbol{O}_i^{(K)\mathsf{T}}(\boldsymbol{y}^{(K)}-\boldsymbol{O}^{(K)}z)\right|}{\left\|\boldsymbol{O}_i^{(K)}\right\|}$

4: Update support $\mathcal{S} \leftarrow \mathcal{S} \cup \mathcal{I}^*$

5: Determine $z_{\mathcal{S}} \leftarrow \boldsymbol{O}_{\mathcal{S}}^{(K)\dagger}\boldsymbol{y}^{(K)}$ and $z_{\mathcal{S}^{\mathsf{C}}} \leftarrow \boldsymbol{0}$

6: Determine new support $\mathcal{S} \leftarrow \underset{\substack{\mathcal{I}\subseteq\{1,2,\ldots,n\}\\|\mathcal{I}|\leq s}}{\arg\max} \sum_{i\in\mathcal{I}} |z_i|$

7: Update the estimate $z_{\mathcal{S}^{\mathsf{C}}} \leftarrow \boldsymbol{0}$

8: **until** $\boldsymbol{y}^{(K)} = \boldsymbol{O}^{(K)}z$

Output: Initial state $\boldsymbol{x}_0 = z$

We note that the optimization problems solved in the orthogonal matching pursuit (Step 3) and compressive sampling matching pursuit (Steps 3 and 6) algorithms need not be unique. In that case, we follow a predefined rule, for example, the lexicographic order.

2.3.3 *Thresholding-Based Methods*

The family of thresholding algorithms relies on a hard thresholding operator that keeps a few largest absolute entries and sets the other ones to zero. A basic thresholding algorithm determines the support set $\mathcal{S}$ as the indices of the s largest absolute entries of $\boldsymbol{O}^{(K)\mathsf{T}}\boldsymbol{y}^{(K)}$ and computes $\boldsymbol{x}_0$ as

$$[\boldsymbol{x}_0]_{\mathcal{S}} = \boldsymbol{O}_{\mathcal{S}}^{(K)\dagger}\boldsymbol{y}^{(K)} \quad \text{and} \quad [\boldsymbol{x}_0]_{\mathcal{S}^{\mathsf{C}}} = \boldsymbol{0}. \tag{2.15}$$

However, this approach works only if the true support coincides with the estimated support. An alternative approach that can circumvent this issue is to solve $\boldsymbol{O}^{(K)\mathsf{T}}\boldsymbol{y}^{(K)} = \boldsymbol{O}^{(K)\mathsf{T}}\boldsymbol{O}^{(K)}\boldsymbol{x}_0$ instead of (2.10) using fixed point equation,

$$z = z + O^{(K)\mathsf{T}} \left(y^{(K)} - O^{(K)} z \right). \tag{2.16}$$

The fixed point iterations are combined with a thresholding operator to retain s largest absolute entries of the solution. The resulting algorithm, called iterative hard thresholding, is summarized in Algorithm 2.4.

Algorithm 2.4 Sparse initial state estimation via iterative hard thresholding

---

Input: Outputs $\{y_k \in \mathbb{R}^p\}_{k=0}^{K-1}$, control inputs $\{u_k \in \mathbb{R}^m\}_{k=1}^{K-1}$ linear dynamical system ($A \in \mathbb{R}^{n\times n}$, $B \in \mathbb{R}^{n\times m}$, $C_{(k)} \in \mathbb{R}^{p\times m}$, $k = 0, 1, \ldots, K-1$), sparsity level s

Initialization: $z = \mathbf{0}$

1: Compute $y^{(K)}$ and $O^{(K)}$ using (2.7) and (2.9)
2: **repeat**
3: Update estimate $z \leftarrow z + O^{(K)\mathsf{T}} \left(y^{(K)} - O^{(K)} z \right)$
4: Determine new support $\mathcal{S} \leftarrow \underset{\substack{\mathcal{I}\subseteq\{1,2,\ldots,n\} \\ |\mathcal{I}|\leq s}}{\arg\max} \sum_{i\in\mathcal{I}} |z_i|$
5: Update the estimate $z_{\mathcal{S}^{\complement}} \leftarrow \mathbf{0}$
6: **until** $y^{(K)} = O^{(K)} z$

Output: Initial state $x_0 = z$

The iterative hard thresholding algorithm can be further improved by recomputing the estimate using least squares after updating the support. This idea leads to the hard thresholding pursuit algorithm, as given in Algorithm 2.5. Here, the extra pseudoinverse in Step 5 increases the computational complexity per iteration of the algorithm while offering a better reconstruction accuracy.

Algorithm 2.5 Sparse initial state estimation via hard thresholding pursuit

---

Input: Outputs $\{y_k \in \mathbb{R}^p\}_{k=0}^{K-1}$, control inputs $\{u_k \in \mathbb{R}^m\}_{k=1}^{K-1}$ linear dynamical system ($A \in \mathbb{R}^{n\times n}$, $B \in \mathbb{R}^{n\times m}$, $C_{(k)} \in \mathbb{R}^{p\times m}$, $k = 0, 1, \ldots, K-1$), sparsity level s

Initialization: $z = \mathbf{0}$

1: Compute $y^{(K)}$ and $O^{(K)}$ using (2.7) and (2.9)
2: **repeat**
3: Update estimate $z \leftarrow z + O^{(K)\mathsf{T}} \left(y^{(K)} - O^{(K)} z \right)$
4: Determine new support $\mathcal{S} \leftarrow \underset{\substack{\mathcal{I}\subseteq\{1,2,\ldots,n\} \\ |\mathcal{I}|\leq s}}{\arg\max} \sum_{i\in\mathcal{I}} |z_i|$
5: Determine $z_{\mathcal{S}} \leftarrow O_{\mathcal{S}}^{(K)\dagger} y^{(K)}$ and $z_{\mathcal{S}^{\complement}} \leftarrow \mathbf{0}$
6: **until** $y^{(K)} = O^{(K)} z$

Output: Initial state $x_0 = z$

We emphasize that the common idea behind all the thresholding algorithms is the intuition that the support is recovered by approximating the inverse of $O^{(K)}$ with its transpose $O^{(K)\mathsf{T}}$.

2.3.4 Bayesian Approach

Within the Bayesian framework, the sparsity constraint is integrated into the estimation problem by an appropriate choice of prior distribution for the sparse vector and then outputting the posterior mean of the vector as an estimate of the sparse vector. A good choice of prior distribution induces an inherent bias towards sparse solutions, which influences the posterior probability distribution and subsequently leads to a sparse initial state estimate. One common choice to promote sparsity is the Gaussian prior with unknown hyperparameters (e.g., the variances of the entries), which leads to an algorithm called sparse Bayesian learning.

For ease of exposition, we consider the estimation of the sparse initial state from noisy outputs, i.e., our model is

$$\boldsymbol{y}^{(K)} = \boldsymbol{O}^{(K)}\boldsymbol{x}_0 + \boldsymbol{w}, \tag{2.17}$$

where $\boldsymbol{w}$ is Gaussian distributed noise with zero mean and covariance matrix $\sigma^2\boldsymbol{I}$. We also assume that $\sigma^2 > 0$ is known. When σ^2 is set close to zero, we obtain the noiseless case in (2.10). Hence, we get the distribution of the noisy output given the initial state as

$$\boldsymbol{y}^{(K)}|\boldsymbol{x}_0;\sigma^2 \sim \mathcal{N}(\boldsymbol{O}^{(K)}\boldsymbol{x}_0, \sigma^2\boldsymbol{I}), \tag{2.18}$$

where $\mathcal{N}(\mu, \nu)$ denotes the Gaussian distribution with mean μ and covariance ν. As mentioned above, we choose a Gaussian prior on $\boldsymbol{x}_0$ with zero mean and an unknown diagonal covariance matrix, $\boldsymbol{\Gamma} \in \mathbb{R}^{n\times n}$,

$$\boldsymbol{\Gamma} = \begin{bmatrix} \gamma_1 & 0 & \dots & \\ 0 & \gamma_2 & & \\ \vdots & & \ddots & \\ & \dots & 0 & \gamma_n \end{bmatrix} = \operatorname{diag}\{\boldsymbol{\gamma}\}. \tag{2.19}$$

Here, $\boldsymbol{\gamma} \in \mathbb{R}^n_+$ is the known hyperparameter vector. As a result, we arrive at

$$\boldsymbol{x}_0|\boldsymbol{\gamma} \sim \mathcal{N}(\boldsymbol{0}, \operatorname{diag}\{\boldsymbol{\gamma}\}). \tag{2.20}$$

In this Bayesian approach, we note that the choice of the Gaussian prior on $\boldsymbol{x}_0$ is fictitious; it is only used to derive the algorithm. The true initial vector can be deterministic or drawn from a different distribution. The Gaussian prior helps us to derive a tractable algorithm. Since we are only concerned about a single vector $\boldsymbol{x}_0$, the true underlying distribution has little effect on the recovery performance.

If $\boldsymbol{\gamma}$ is known to us, we can recover $\boldsymbol{x}_0$ using its posterior distribution given the output $\boldsymbol{y}^{(K)}$. Specifically, we see that the distribution of $\boldsymbol{x}_0$ conditioned on $\boldsymbol{y}^{(K)}$ is Gaussian with mean $\boldsymbol{\mu}$ and covariance $\boldsymbol{\Sigma}$ given by

$$\boldsymbol{\mu} = \sigma^{-2}\boldsymbol{\Sigma}\boldsymbol{O}^{(K)\mathsf{T}}\boldsymbol{y}^{(K)} \tag{2.21}$$

$$\boldsymbol{\Sigma} = \left(\sigma^{-2}\boldsymbol{O}^{(K)\mathsf{T}}\boldsymbol{O}^{(K)} + \operatorname{diag}\{\boldsymbol{\gamma}\}^{-1}\right)^{-1}. \tag{2.22}$$

So, the estimate of $\boldsymbol{x}_0$ given $\boldsymbol{\gamma}$ is $\boldsymbol{\mu}$. Further, to estimate the hyperparameters $\boldsymbol{\gamma}$ using $\boldsymbol{y}^{(K)}$, we maximize the likelihood of the output from (2.17) and (2.20), i.e., we seek $\boldsymbol{\gamma}$ that maximizes

$$p(\boldsymbol{y}^{(K)}; \boldsymbol{\gamma}, \sigma^2) \sim \mathcal{N}\left(\boldsymbol{0}, \sigma^2\boldsymbol{I} + \boldsymbol{O}^{(K)}\operatorname{diag}\{\boldsymbol{\gamma}\}\,\boldsymbol{O}^{(K)\mathsf{T}}\right). \tag{2.23}$$

Unfortunately, maximizing the above likelihood does not yield a closed-form solution, and we resort to an iterative solution via the expectation–maximization algorithm. Every iteration of the expectation-maximization algorithm finds the estimate of $\boldsymbol{\gamma}$ that maximizes the expected value of the log-likelihood function of $\boldsymbol{\gamma}$, given by $\log p(\boldsymbol{x}_0, \boldsymbol{y}^{(K)}; \boldsymbol{\gamma}, \sigma^2)$, with respect to the current conditional distribution of $\boldsymbol{x}_0$ given $\boldsymbol{y}^{(K)}$ and the current estimates of the parameter $\boldsymbol{\gamma}'$. Therefore, an iteration of the expectation-maximization algorithm can be summarized as

$$\boldsymbol{\gamma}^* = \underset{\boldsymbol{\gamma}\in\mathbb{R}_+^n}{\arg\max}\ \mathbb{E}_{\boldsymbol{x}_0|\boldsymbol{y}^{(K)};\boldsymbol{\gamma}',\sigma^2}\left\{\log p(\boldsymbol{x}_0, \boldsymbol{y}^{(K)}; \operatorname{diag}\{\boldsymbol{\gamma}\}, \sigma^2)\right\}. \tag{2.24}$$

Interestingly, the above optimization problem has a closed-form solution given by

$$\gamma_i^* = \mu_i^2 + \Sigma_{ii}. \tag{2.25}$$

The resulting algorithm is summarized in Algorithm 2.6. A detailed derivation of the sparse Bayesian learning algorithm is presented in Appendix A.2.

Algorithm 2.6 Sparse initial state estimation via sparse Bayesian learning

Input: Outputs $\{\boldsymbol{y}_k \in \mathbb{R}^p\}_{k=0}^{K-1}$, control inputs $\{\boldsymbol{u}_k \in \mathbb{R}^m\}_{k=1}^{K-1}$ linear dynamical system ($\boldsymbol{A} \in \mathbb{R}^{n\times n}$, $\boldsymbol{B} \in \mathbb{R}^{n\times m}$, $\boldsymbol{C}_{(k)} \in \mathbb{R}^{p\times m}$, $k = 0, 1, \ldots, K-1$),
Initialization: $\boldsymbol{\gamma} = \boldsymbol{1}$ and σ^2 to a small positive value
1: Compute $\boldsymbol{y}^{(K)}$ and $\boldsymbol{O}^{(K)}$ using (2.7) and (2.9)
2: **repeat**
3: Update estimate $\boldsymbol{\mu}$ and $\boldsymbol{\Sigma}$ using (2.21) and (2.22).
4: Determine new hyperparameter $\gamma_i = \mu_i^2 + \Sigma_{ii}$, for $i = 1, 2, \ldots, n$.
5: **until** $\boldsymbol{y}^{(K)} = \boldsymbol{O}^{(K)}\boldsymbol{\mu}$
Output: Initial state $\boldsymbol{x}_0 = \boldsymbol{\mu}$

We emphasize that the stopping criterion in Algorithms 2.2–2.6 is met only when we find an initial state that precisely fits the output vector, e.g., $\boldsymbol{y}^{(K)} = \boldsymbol{O}^{(K)}\boldsymbol{\mu}$. For iterative algorithms, this criterion is hard to achieve in practice because of numerical errors and asymptotic convergence. Therefore, for practical purposes, we use

$\left\| \boldsymbol{y}^{(K)} - \boldsymbol{O}^{(K)} \boldsymbol{\mu} \right\| \leq \eta$ as the stopping criterion, where $\eta > 0$ is a small error tolerance.

2.4 Summary

- If the initial state of a linear dynamical system is sparse, it can be estimated with far fewer observations than its dimension.
- The signal processing technique to recover a sparse vector from an underdetermined linear system is called compressed sensing.
- The compressed sensing algorithms come in various flavors, such as convex optimization, greedy approaches, thresholding-based, and Bayesian approaches.

Notes

The specifics of various algorithms in compressed sensing, such as convex optimization, greedy, and thresholding-based algorithms, are detailed in Foucart and Rauhut (2013) and its cited references. Further, within the realm of convex optimization, additional variants include basis pursuit denoising Chen et al. (2001), least absolute shrinkage and selection operator (LASSO) Tibshirani (1996), Efron et al. (2004), and the Dantzig selector Candes and Tao (2007). We also excluded a discussion on the greedy methods, known as matching pursuit Mallat and Zhang (1993) and regularized orthogonal matching pursuit Needell and Vershynin (2009), and the threshold-based method called iterative soft thresholding Bredies and Lorenz (2008), Blumensath and Davies (2009). Another algorithmic category utilizes heavy-tailed priors to encourage sparse solutions, resulting in nonconvex optimization problems. Examples encompass the FOcal Underdetermined System Solver Gorodnitsky and Rao (1997), Bayesian compressive sensing using Laplace priors Babacan et al. (2009), as well as sampling-based Monte Carlo approaches George and McCulloch (1997), George and McCulloch (1993), Olshausen and Millman (1999). Additionally, iterative reweighted ℓ_1 and ℓ_2 methods aim to minimize nonconvex cost functions through successive convex approximations. Furthermore, techniques for sparse recovery also comprise approximate message passing Donoho et al. (2009), deep learning approaches Mousavi et al. (2015), and generative models Bora et al. (2017).

References

Babacan SD, Molina R, Katsaggelos AK (2009) Bayesian compressive sensing using Laplace priors. IEEE Trans Image Process 19(1):53–63

Blumensath T, Davies ME (2009) Iterative hard thresholding for compressed sensing. Appl Comput Harmon Anal 27(3):265–274

Bora A, Jalal A, Price E, Dimakis AG (2017) Compressed sensing using generative models. In: Proceedings of international conference on machine learning, PMLR, pp 537–546

Bredies K, Lorenz DA (2008) Linear convergence of iterative soft-thresholding. J Fourier Anal Appl 14:813–837

Candes E, Tao T (2007) The Dantzig selector: statistical estimation when p is much larger than n. Ann Stat 35(6):2313–2351

Chen SS, Donoho DL, Saunders MA (2001) Atomic decomposition by basis pursuit. SIAM Rev 43(1):129–159

Donoho DL, Maleki A, Montanari A (2009) Message-passing algorithms for compressed sensing. Proc Natl Acad Sci 106(45):18914–18919

Efron B, Hastie T, Johnstone I, Tibshirani R (2004) Least angle regression. Ann Stat 32(2):407–499

Foucart S, Rauhut H (2013) An invitation to compressive sensing. Springer

George EI, McCulloch RE (1997) Approaches for Bayesian variable selection. Statistica sinica 339–373

George EI, McCulloch RE (1993) Variable selection via Gibbs sampling. J Am Stat Assoc 88(423):881–889

Gorodnitsky IF, Rao BD (1997) Sparse signal reconstruction from limited data using FOCUSS: a re-weighted minimum norm algorithm. IEEE Trans Signal Process 45(3):600–616

Mallat SG, Zhang Z (1993) Matching pursuits with time-frequency dictionaries. IEEE Trans Signal Process 41(12):3397–3415

Mousavi A, Patel AB, Baraniuk RG (2015) A deep learning approach to structured signal recovery. In: Proceedings of Allerton conference on communication, control, and computing, pp 1336–1343

Needell D, Vershynin R (2009) Uniform uncertainty principle and signal recovery via regularized orthogonal matching pursuit. Found Comput Math 9:317–334

Olshausen B, Millman K (1999) Learning sparse codes with a mixture-of-gaussians prior. Adv Neural Inf Process Syst 12

Tibshirani R (1996) Regression shrinkage and selection via the LASSO. J R Stat Soc Ser B Stat Methodol 58(1):267–288

Chapter 3
Observability of Sparse Initial State

Abstract In this chapter, we establish theoretical guarantees for the recovery of the sparse initial state of a linear dynamical system using the algorithms presented in Chapter 2. We rely on the concept of the restricted isometry property and prove that if the initial state vector has a sparse representation, the number of necessary observations can be significantly reduced through random projections. Our analysis provides sufficient conditions for the observability matrix to possess the restricted isometry property, thereby ensuring the system's observability. These results depend solely on the characteristics of the system's transfer and observation matrices, employing tools from probability theory and compressed sensing.

Keywords Restricted isometry property · subGaussian matrices · Stochastic linear dynamical systems · Measurement bound · Singular value ratio · Sparse observability index

3.1 Sparse Observability

Observability is an important property for characterizing the behavior of a linear dynamical system. To begin our discussion, we recall the state space model of the linear dynamical system ($\boldsymbol{A} \in \mathbb{R}^{n\times n}$, $\boldsymbol{B} \in \mathbb{R}^{n\times m}$, $\boldsymbol{C}_{(k)} \in \mathbb{R}^{p\times m},\ k = 0, 1, \ldots$), with state $\boldsymbol{x}_k \in \mathbb{R}^n$, input $\boldsymbol{u}_k \in \mathbb{R}^m$, and output $\boldsymbol{y}_k$,

$$\boldsymbol{x}_k = \boldsymbol{A}\boldsymbol{x}_{k-1} + \boldsymbol{B}\boldsymbol{u}_k, \qquad \text{and} \qquad \boldsymbol{y}_k = \boldsymbol{C}_{(k)}\boldsymbol{x}_k, \tag{3.1}$$

for discrete-time index k. If we can estimate the state $\boldsymbol{x}_k$ of a linear dynamical system for any k using its observations with the knowledge of control inputs, the system is observable. Also, if the initial state $\boldsymbol{x}_0$ is known, then the state $\boldsymbol{x}_k$ at all k can be determined using the state dynamics. So, we focus on the initial state estimation.

We recall the sparse initial state estimation problem from Chapter 2,

$$\boldsymbol{y}^{(K)} = \boldsymbol{O}^{(K)}\boldsymbol{x}_0, \tag{3.2}$$

G. Joseph and C. R. Murthy, *Sparsity-Constrained Linear Dynamical Systems*,
Springer Tracts in Electrical and Electronics Engineering,
https://doi.org/10.1007/978-981-97-7090-8_3

where the measurements $\boldsymbol{y}^{(K)} \in \mathbb{R}^{pK}$ and measurement matrix $\boldsymbol{O}^{(K)} \in \mathbb{R}^{pK \times n}$ are

$$\boldsymbol{y}^{(K)} = \begin{bmatrix} y_0 \\ y_1 \\ \vdots \\ y_{K-1} \end{bmatrix} - \begin{bmatrix} \mathbf{0} & & & \\ \boldsymbol{C}_{(1)}\boldsymbol{B} & & & \\ \boldsymbol{C}_{(2)}\boldsymbol{A}\boldsymbol{B} & \boldsymbol{C}_{(2)}\boldsymbol{B} & & \\ \vdots & \vdots & \ddots & \\ \boldsymbol{C}_{(K-1)}\boldsymbol{A}^{K-2}\boldsymbol{B} & \boldsymbol{C}_{(K-1)}\boldsymbol{A}^{K-3}\boldsymbol{B} & \dots & \boldsymbol{C}_{(K-1)}\boldsymbol{B} \end{bmatrix} \begin{bmatrix} u_1 \\ u_2 \\ \vdots \\ u_{K-1} \end{bmatrix} \tag{3.3}$$

$$\boldsymbol{O}^{(K)} = \begin{bmatrix} \boldsymbol{C}_{(0)} \\ \boldsymbol{C}_{(1)}\boldsymbol{A} \\ \vdots \\ \boldsymbol{C}_{(K-1)}\boldsymbol{A}^{K-1} \end{bmatrix}. \tag{3.4}$$

A linear dynamical system is said to be observable if, for any initial state $\boldsymbol{x}_0$, there exists an integer $K < \infty$ such that $\boldsymbol{x}_0$ can be determined uniquely from the outputs $\{y_k\}_{k=0}^{K-1}$, the transfer matrix $\boldsymbol{A}$, and the observation matrices $\{\boldsymbol{C}_{(k)}\}_{k=0}^{K-1}$. In other words, a system is observable if the initial state can be estimated from the system of Eqs. (3.2) for some finite $K > 0$. The notation of observability was introduced by R. E. Kalman and he also introduced the Kalman rank test to check whether a given system is observable.

Theorem 3.1 *The linear dynamical system* ($\boldsymbol{A} \in \mathbb{R}^{n\times n}, \boldsymbol{B} \in \mathbb{R}^{n\times m}, \boldsymbol{C}_{(k)} \in \mathbb{R}^{p\times m}$, $k = 0, 1, \dots$) *in* (3.1) *is said to be observable if and only if it satisfies the Kalman rank condition given by*

$$\operatorname{rank}\left\{\boldsymbol{O}^{(K)}\right\} = n, \tag{3.5}$$

for some finite $K < \infty$. *Specifically, the system is observable only if*

$$pK \geq n. \tag{3.6}$$

We next turn to the sparse initial state estimation problem. In this chapter, we investigate how the additional sparsity information helps relax the bound in (3.6). The analysis of recovery algorithms in Chapter 2 usually involves a quantity that measures the suitability of the measurement matrix $\boldsymbol{O}^{(K)}$ for recovering sparse vectors. In this chapter, we focus on a measure of the quality of a measurement matrix called the restricted isometry property. It ensures the success of several sparse recovery algorithms, including ℓ_1-minimzation, orthogonal matching pursuit, compressive sampling matching pursuit, and iterative hard thresholding, presented in Chapter 2.

3.2 Restricted Isometry Property

This section introduces the concept of restricted isometry property and the theoretical guarantees on compressed sensing algorithms based on restricted isometry property.

Definition 3.1 The sth restricted isometry constant $\delta_s(\boldsymbol{M})$ of a matrix $\boldsymbol{M} \in \mathbb{R}^{M \times N}$ is the smallest $\delta \geq 0$ such that

$$1 - \delta \leq \|\boldsymbol{M}\boldsymbol{x}\|^2 \leq 1 + \delta, \tag{3.7}$$

for all s-sparse vectors $\boldsymbol{x} \in \mathbb{R}^N$ with $\|\boldsymbol{x}\| = 1$.

A small restricted isometry constant refers to matrices that are nearly orthonormal, at least when operating on a sparse vector. Naturally, a smaller restricted isometry constant of the measurement matrix favors the graceful recovery of sparse vectors. We next state (without proof) the available guarantees for successful sparse vector recovery via different compressed sensing algorithms for measurement matrices with small restricted isometry constants.

3.2.1 Algorithm Guarantees

We consider a general sparse recovery problem that seeks a sparse solution to an underdetermined system of equations,

$$\boldsymbol{y} = \boldsymbol{M}\boldsymbol{x}, \tag{3.8}$$

where $\boldsymbol{y} \in \mathbb{R}^M$ is the measurement vector, $\boldsymbol{M} \in \mathbb{R}^{M \times N}$ is the measurement matrix, and $\boldsymbol{x} \in \mathbb{R}^N$ is the unknown sparse vector. The sufficient condition in this section requires the restricted isometry constant of the measurement matrix to satisfy an upper bound on δ_{qs}, for some $q > 0$. The values of q and the upper bound depend on the specific compressive sensing algorithm. We summarize the results in the following.

Theorem 3.2 *Suppose the* $2s$*th of the restricted isometry constant of a matrix* $\boldsymbol{M} \in \mathbb{R}^{M \times N}$ *satisfies* $\delta_{2s}(\boldsymbol{M}) < 1/3$. *Then, for every* s*-sparse vector* $\boldsymbol{x} \in \mathbb{R}^N$, *the basis pursuit algorithm returns* $\boldsymbol{x}$ *satisfying* $\boldsymbol{y} = \boldsymbol{M}\boldsymbol{x}$.

Theorem 3.3 *Suppose the* $4s$*th of the restricted isometry constant of a matrix* $\boldsymbol{M} \in \mathbb{R}^{M \times N}$ *satisfies* $\delta_{4s}(\boldsymbol{M}) < \sqrt{\sqrt{11/3} - 1}/2$. *Then, for every* s*-sparse vector* $\boldsymbol{x} \in \mathbb{R}^N$, *the compressive sampling matching pursuit algorithm converges to* $\boldsymbol{x}$ *satisfying* $\boldsymbol{y} = \boldsymbol{M}\boldsymbol{x}$.

Theorem 3.4 *Suppose the* $3s$*th of the restricted isometry constant of a matrix* $\boldsymbol{M} \in \mathbb{R}^{M \times N}$ *satisfies* $\delta_{3s}(\boldsymbol{M}) < 1/2$. *Then, for every* s*-sparse vector* $\boldsymbol{x} \in \mathbb{R}^N$, *the iterative hard thresholding algorithm converges to* $\boldsymbol{x}$ *satisfying* $\boldsymbol{y} = \boldsymbol{M}\boldsymbol{x}$.

To summarize, the sparse initial state estimation from (3.2) is successful if the restricted isometry constant of $\boldsymbol{O}^{(K)}$ is small enough. Therefore, in the following, we analyze the restricted isometry property of $\boldsymbol{O}^{(K)}$ to establish the observability of the sparse initial state.

3.2.2 Difficulty of Analyzing Restricted Isometry Property

As mentioned, the restricted isometry property is an excellent measure for analyzing several compressed sensing algorithms. However, computing the restricted isometry constant of a given matrix is itself an NP-hard problem. It is also difficult to approximate the restricted isometry constant of a matrix, making this measure "useless". Moreover, it is infamously challenging to find a deterministic matrix that satisfies a given upper bound on the restricted isometry constant. Here, a deterministic matrix refers to an explicit matrix construction or polynomial algorithms to construct such a matrix. To add to the difficulty, for the initial state estimation problem, the measurement matrix, which is the observability matrix, has a special structure that depends on the state and observation matrices, owing to the state space model. Hence, finding a good choice of state and observation matrices that guarantees sparse initial state recoverability is a difficult and open problem.

Luckily, when the measurement matrix is random, we can leverage powerful analytical techniques rooted in probability theory to assess the restricted isometry property. Specifically, it is known in the compressed sensing literature that random Gaussian, Bernoulli, and partial Fourier matrices fulfill the restricted isometry property with an exponentially high probability (in the matrix dimension). Interestingly, it is also known that the number of measurements required to recover a sparse vector scales linearly in the sparsity level and logarithmically in its dimension. Hence, we look at stochastic linear dynamical systems and analyze their observability under sparsity constraints on the initial state.

Although randomness in the measurement helps the analysis of the restricted isometry property, the analysis of the observability matrix of a stochastic linear dynamical system is different from the analysis of a general random measurement matrix. The major bottleneck in the observability analysis is the structure of the observability matrix that involves different powers of $\boldsymbol{A}$. Hence, the analysis of the structured random matrix is more intricate than the typical compressed sensing measurement matrix, as seen in the following sections.

3.3 Sparse Observability with Sub-Gaussian Observation Matrices

We consider a stochastic system when the observation matrix is a sub-Gaussian random matrix and look at the restricted isometry property of the observability matrix $\boldsymbol{O}^{(K)}$ in (3.4). The restricted isometry property analysis combined with Theorems 3.2–3.4 provides sufficient conditions for sparse observability. For this analysis, we assume that the observation matrices are sub-Gaussian and the state matrix is bounded. More precisely, the following are the assumptions used for our analysis:

Assumption 3.1 *The observation matrices are independent and identically distributed sub-Gaussian random matrices defined in Definition 3.2.*

Here, a sub-Gaussian random matrix is defined as follows:

Definition 3.2 A random variable Z is said to be sub-Gaussian with parameters $\beta, \kappa > 0$ if, for any $t > 0$, it satisfies

$$\mathbb{P}\{|Z| \geq t\} \leq \beta \exp(-\kappa t^2). \tag{3.9}$$

Also, if the entries of the matrix are independent, zero mean, unit variance sub-Gaussian random variables with the same parameters $\beta, \kappa > 0$, then the matrix is said to be a sub-Gaussian random matrix.

The sub-Gaussian distribution has strong tail decay so that the tails are exponentially bounded. Some example distributions are Gaussian, Rademacher, uniform distribution over a bounded interval, etc.

As discussed in Sect. 3.2.2, the sub-Gaussian assumption is motivated by the small restricted isometry constants of sub-Gaussian matrices and the richness of the associated toolset. The second assumption is on the largest singular value of the state matrix.

Assumption 3.2 *The largest singular value of the state matrix* $\boldsymbol{A}$ *is 1.*

This assumption is not a strong one and is imposed only for the simplicity of the analysis. Suppose $\xi_{\max} \neq 1$ is the largest singular value of the state matrix $\boldsymbol{A}$. Then, $\tilde{\boldsymbol{A}} = \boldsymbol{A}/\xi_{\max}$ has 1 as its largest singular value. We note that $\xi_{\max} > 0$ because the largest singular is zero if and only if $\boldsymbol{A} = \boldsymbol{0}$. So, $1/\xi_{\max} < \infty$ and using (3.4), the system of equations in (3.2) is equivalent to

$$\begin{bmatrix} \boldsymbol{I} & & & \\ & \xi_{\max}\boldsymbol{I} & & \\ & & \ddots & \\ & & & \xi_{\max}^{K-1}\boldsymbol{I} \end{bmatrix}^{-1} \boldsymbol{y}^{(K)} = \begin{bmatrix} \boldsymbol{I} & & & \\ & \xi_{\max}\boldsymbol{I} & & \\ & & \ddots & \\ & & & \xi_{\max}^{K-1}\boldsymbol{I} \end{bmatrix}^{-1} \boldsymbol{O}^{(K)}\boldsymbol{x}_0 \tag{3.10}$$

$$= \begin{bmatrix} I & & & \\ & \xi_{\max} I & & \\ & & \ddots & \\ & & & \xi_{\max}^{K-1} I \end{bmatrix}^{-1} \begin{bmatrix} C_{(0)} \\ C_{(1)} A \\ \vdots \\ C_{(K-1)} A^{K-1} \end{bmatrix} x_0 \tag{3.11}$$

$$= \begin{bmatrix} C_{(0)} \\ C_{(1)} A / \xi_{\max} \\ \vdots \\ C_{(K-1)} A^{K-1} / \xi_{\max}^{K-1} \end{bmatrix} x_0 \tag{3.12}$$

$$= \begin{bmatrix} C_{(0)} \\ C_{(1)} \tilde{A} \\ \vdots \\ C_{(K-1)} \tilde{A}^{K-1} \end{bmatrix} x_0. \tag{3.13}$$

If the system with the modified state matrix $(\tilde{A}, B, C_k,\ k = 0, 1, \ldots, K-1)$ is observable, the linear system $(A, B, C_k,\ k = 0, 1, \ldots, K-1)$ is also observable. Hence, without loss of generality, we make Assumption 3.2.

The main result of the section is as follows.

Theorem 3.5 *Consider the linear dynamical system* $(A \in \mathbb{R}^{n\times n}, B \in \mathbb{R}^{n\times m}, C_{(k)} \in \mathbb{R}^{p\times m},\ k = 0, 1, \ldots, K-1)$*. Let* δ *be such that*

$$c(1 - \xi_{\min}^{2(K-1)}) < \delta < 1, \tag{3.14}$$

where $\xi_{\min} > 0$ *is the smallest singular value of* A*. Then, under Assumptions 3.1 and 3.2, there exist constants* $c > \frac{2}{\sqrt{3}}$ *and* $C_1, C_2 > 0$ *such that the restricted isometry constant of* $\frac{1}{\sqrt{pK}} O^{(K)}$ *in* (3.4) *satisfies* $\delta_s \leq \delta$ *with probability at least* $1 - \epsilon$ *if, for* $\epsilon \in (0, 1)$,

$$pK\left[\delta - c(1 - \xi_{\min}^{2(K-1)})\right]^2 \geq C_1 s \log(en/s) + C_2 \log(2\epsilon^{-1}). \tag{3.15}$$

The proof of the theorem is based on advanced tools from probability theory outlined in Appendix A.3. Using these tools, we next present two intermediate results used to prove Theorem 3.5.

Lemma 3.1 *For any support set* $\mathcal{S} \subset \{1, 2, \ldots, n\}$*, let the unit ball of vectors supported on* $\mathcal{S}$ *be*

$$\mathbb{B}_{\mathcal{S}} = \{x \in \mathbb{R}^n : \|x\| = 1 \ \textit{and}\ \operatorname{supp}(x) \subseteq \mathcal{S}\}. \tag{3.16}$$

Then, for any $0 < \rho < 1/2$*, there exists a finite set of sparse vectors* $\mathcal{Z}_{\mathcal{S}} \subset \mathbb{B}_{\mathcal{S}}$ *with* $|\mathcal{Z}_{\mathcal{S}}| \leq \left(1 + \frac{2}{\rho}\right)^s$*, such that, for any matrix* $M \in \mathbb{R}^{M\times n}$,

$$\max_{x\in\mathbb{B}_S} \left|\|Mx\|^2 - \|x\|^2\right| \leq \frac{1}{1-2\rho} \max_{z\in\mathcal{Z}_S} \left|\|Mz\|^2 - \|z\|^2\right|. \tag{3.17}$$

Proof Invoking the covering number-based arguments (see Proposition A.2), there exists a set of balls that cover the unit ball $\mathbb{B}_S$. Let $\mathcal{Z}_S \subset \mathbb{B}_S$ be such that

$$|\mathcal{Z}_S| \leq \left(1+\frac{2}{\rho}\right)^s \quad \text{and} \quad \mathbb{B}_S \subseteq \bigcup_{z\in\mathcal{Z}_S} \{z' \in \mathbb{R}^n : \ \|z-z'\| \leq \rho\}. \tag{3.18}$$

Then, for any vector $x \in \mathbb{B}_S$, there exists a vector $z \in \mathcal{Z}_S$ such that

$$\|x-z\| \leq \rho. \tag{3.19}$$

Now, to prove the result, we next verify (3.17). To this end, we note that

$$\begin{aligned}
\left|\|Mx\|^2 - \|x\|^2\right| &= \left|x^\mathsf{T}\left(M^\mathsf{T}M - I\right)x\right| &(3.20)\\
&= \left|z^\mathsf{T}\left(M^\mathsf{T}M - I\right)z + (x+z)^\mathsf{T}\left(M^\mathsf{T}M - I\right)(x-z)\right| &(3.21)\\
&\leq \left|z^\mathsf{T}\left(M^\mathsf{T}M - I\right)z\right| + \left|(x+z)^\mathsf{T}\left(M^\mathsf{T}M - I\right)(x-z)\right| &(3.22)\\
&= \left|\|Mz\|^2 - \|z\|^2\right| + \left|(x+z)^\mathsf{T}\left(M^\mathsf{T}M - I\right)(x-z)\right|. &(3.23)
\end{aligned}$$

Since $x, z \in \mathbb{B}_S$, we simplify the last term here as

$$\begin{aligned}
(x+z)^\mathsf{T}\left(M^\mathsf{T}M - I\right)(x-z) &= (x_S+z_S)^\mathsf{T}\left(M_S^\mathsf{T}M_S - I\right)(x_S - z_S) &(3.24)\\
&\leq \|x+z\| \left\|M_S^\mathsf{T}M_S - I\right\| \|x-z\| &(3.25)\\
&\leq (\|x\| + \|z\|) \left\|M_S^\mathsf{T}M_S - I\right\| \|x-z\| &(3.26)\\
&\leq 2\rho \left\|M_S^\mathsf{T}M_S - I\right\| &(3.27)\\
&\leq 2\rho \max_{x'\in\mathbb{R}^{|S|}:\|x'\|=1} \left|x'^\mathsf{T}\left(M_S^\mathsf{T}M_S - I\right)x'\right| &(3.28)\\
&= 2\rho \max_{x\in\mathbb{B}_S} \left|x^\mathsf{T}\left(M^\mathsf{T}M - I\right)x\right| &(3.29)\\
&= 2\rho \max_{x\in\mathbb{B}_S} \left|\|Mx\|^2 - \|x\|^2\right|, &(3.30)
\end{aligned}$$

where (3.27) uses the fact that $\|x\| = \|z\| = 1$ and (3.19). Combining (3.23) and (3.30) leads to

$$\max_{x\in\mathbb{B}_S} \left|\|Mx\|^2 - \|x\|^2\right| \leq \max_{z\in\mathcal{Z}_S} \left|\|Mz\|^2 - \|z\|^2\right| + 2\rho \max_{x\in\mathbb{B}_S} \left|\|Mx\|^2 - \|x\|^2\right|. \tag{3.31}$$

Rearranging the above relation, we arrive at (3.17), and the proof is complete. □

Lemma 3.2 *For any matrix $M \in \mathbb{R}^{M\times n}$, suppose any $z \in \mathcal{Z}_S$ in Lemma 3.1 where $S \subseteq \{1, 2, \ldots, n\}$ with $|S| = s$, satisfies*

$$\mathbb{P}\left\{\left|\|\boldsymbol{M}z\|^2 - \|z\|^2\right| \geq t\right\} \leq f(t), \tag{3.32}$$

for some function $f(t) \in [0, 1]$, *then, for any* $0 < \rho < 1/2$, *the restricted isometry constant satisfies*

$$\mathbb{P}(\delta_s \geq \delta) \leq \left[\frac{en}{s}\left(1 + \frac{2}{\rho}\right)\right]^s f(\delta(1 - 2\rho)). \tag{3.33}$$

Proof From Definition 3.1, the relation (3.7) holds only if $\left|\|\boldsymbol{M}\boldsymbol{x}\|^2 - 1\right| \leq \delta$, for all s-sparse vectors $\boldsymbol{x} \in \mathbb{R}^n$ with $\|\boldsymbol{x}\| = 1$, i.e., the restricted isometry constant satisfies

$$\delta_s \geq \max_{\boldsymbol{x}\in\mathbb{B}_{\mathcal{S}}} \left|\|\boldsymbol{M}\boldsymbol{x}\|^2 - 1\right|, \quad \forall \mathcal{S} \subseteq \{1, 2, \ldots, n\} \text{ with } |\mathcal{S}| \leq s, \tag{3.34}$$

where $\mathbb{B}_{\mathcal{S}}$ is a collection of unit norm vectors supported on $\mathcal{S}$ as defined in Lemma 3.1. Therefore, using the union bound, we have

$$\mathbb{P}\{\delta_s \geq \delta\} = \mathbb{P}\left\{\max_{\boldsymbol{x}\in\mathbb{B}_{\mathcal{S}}} \left|\|\boldsymbol{M}\boldsymbol{x}\|^2 - 1\right| \geq \delta, \quad \forall \mathcal{S} \subseteq \{1, 2, \ldots, n\} \text{ with } |\mathcal{S}| = s\right\} \tag{3.35}$$

$$\leq \sum_{\mathcal{S}\subseteq\{1,2,\ldots,n\}:\ |\mathcal{S}|=s} \mathbb{P}\left\{\max_{\boldsymbol{x}\in\mathbb{B}_{\mathcal{S}}} \left|\|\boldsymbol{M}\boldsymbol{x}\|^2 - 1\right| \geq \delta\right\} \tag{3.36}$$

$$\leq \sum_{\mathcal{S}\subseteq\{1,2,\ldots,n\}:\ |\mathcal{S}|=s} \mathbb{P}\left\{\frac{1}{1-2\rho} \max_{z\in\mathcal{Z}_{\mathcal{S}}} \left|\|\boldsymbol{M}z\|^2 - \|z\|^2\right| \geq \delta\right\}, \tag{3.37}$$

for any $0 < \rho < 1/2$ by Lemma 3.1, using its definition of $\mathcal{Z}_{\mathcal{S}}$. We further bound the probability using the union bound over $\mathcal{Z}_{\mathcal{S}}$ to arrive at

$$\mathbb{P}\{\delta_s \geq \delta\} \leq \sum_{\mathcal{S}\subseteq\{1,2,\ldots,n\}:\ |\mathcal{S}|=s} \ \sum_{z\in\mathcal{Z}_{\mathcal{S}}} f(\delta(1 - 2\rho)) \tag{3.38}$$

$$= \binom{n}{s} |\mathcal{Z}_{\mathcal{S}}| \, f(\delta(1 - 2\rho)) \tag{3.39}$$

$$\leq \binom{n}{s}\left(1 + \frac{2}{\rho}\right)^s f(\delta(1 - 2\rho)), \tag{3.40}$$

where we use Lemma 3.1. However, we know that

$$\binom{n}{s} = \frac{n(n-1)\ldots(n-s+1)}{s!} \leq \frac{n^s}{s!} = \frac{n^s}{s^s}\frac{s^s}{s!} \leq \frac{n^s}{s^s}\sum_{l=0}^{\infty}\frac{s^l}{l!} = \frac{n^s}{s^s}e^s. \tag{3.41}$$

Using this upper bound on the binomial coefficient in (3.40), we arrive at the desired result. □

We are ready to prove Theorem 3.5 using Lemmas 3.1 and 3.2.

Proof (of Theorem 3.5) For any $\boldsymbol{x}$ with $\|\boldsymbol{x}\| = 1$, we have

$$\left|\frac{1}{pK}\left\|\boldsymbol{O}^{(K)}\boldsymbol{x}\right\|^2 - 1\right| = \frac{1}{pK}\left|\sum_{k=0}^{K-1}\sum_{i=1}^{p}(X_{ik}^2 - 1)\right|, \tag{3.42}$$

where $X_{ik} = \left[\boldsymbol{C}_i^{(K)\mathrm{T}}\right]^{\mathrm{T}} \boldsymbol{A}^k\boldsymbol{x}$ and $\boldsymbol{C}_i^{(K)\mathrm{T}}$ is the ith row of $\boldsymbol{C}^{(K)}$. Consequently,

$$\mathbb{P}\left\{\left|\frac{1}{pK}\left\|\boldsymbol{O}^{(K)}\boldsymbol{x}\right\|^2 - 1\right| \geq t\right\} = \mathbb{P}\left\{\left|\sum_{k=0}^{K-1}\sum_{i=1}^{p}X_{ik}^2 - 1\right| \geq pKt\right\}. \tag{3.43}$$

By Assumption 3.1 and using the properties of a sub-Gaussian random variable (see Lemma A.2), we infer that X_{ik} is a zero-mean sub-Gaussian random variable with parameter $\beta, \kappa / \left\|\boldsymbol{A}^k\boldsymbol{x}\right\|^2$. Therefore, from Definition 3.2, we know that

$$\mathbb{P}\{|X_{ik}| \geq t\} \leq \beta \exp\left(-\frac{\kappa t^2}{\left\|\boldsymbol{A}^k\boldsymbol{x}\right\|^2}\right) \leq \beta \exp(-\kappa t^2), \tag{3.44}$$

because $\left\|\boldsymbol{A}^k\boldsymbol{x}\right\|^2 \leq \|\boldsymbol{x}\|^2 \leq 1$ by Assumption 3.2. This relation concludes that X_{ik} is a zero-mean sub-Gaussian random variable with parameter β, κ.

Since the square of a sub-Gaussian random variable is subexponential (see Definition A.2 and Lemma A.1), we obtain that $X_{ik}^2 - 1$ is subexponential with parameters that depend only on β, κ. Also, by the independence of the entries of $\boldsymbol{C}^{(K)}$, the random variables $X_{ik}^2 - 1$ are independent and satisfy

$$\mathbb{E}\left\{X_{ik}^2 - 1\right\} = \mathbb{E}\left\{\left(\left[\boldsymbol{C}_i^{(K)\mathrm{T}}\right]^{\mathrm{T}} \boldsymbol{A}^k\boldsymbol{x}\right)^2 - 1\right\} = \left\|\boldsymbol{A}^k\boldsymbol{x}\right\|^2 - 1, \tag{3.45}$$

which follows from the fact that the entries of $\boldsymbol{C}^{(K)}$ have unit variance by Assumption 3.1 and Definition 3.2. Since $\xi_{\min} \leq 1$ is the smallest singular value of $\boldsymbol{A}$ and by Assumption 3.2, we derive that, for any $k \leq K - 1$

$$\xi_{\min}^{2(K-1)} - 1 \leq \xi_{\min}^{2k} - 1 \leq \mathbb{E}\left\{X_{ik}^2 - 1\right\} \leq 0. \tag{3.46}$$

Therefore, invoking the Bernstein-type inequality for non-centered subexponential random variables (see Lemma A.3), we deduce from (3.43) that, for any $t > 1 - \xi_{\min}^{2(K-1)}$, there exist $\tilde{\beta}, \tilde{\kappa} > 0$ that depend only on β, κ such that

$$\mathbb{P}\left\{\left|\frac{1}{pK}\left\|\boldsymbol{O}^{(K)}\boldsymbol{x}\right\|^2-1\right|\geq t\right\}\leq 2\exp\left(-\frac{\tilde{\kappa}^2[pKt-pK(1-\xi_{\min}^{2(K-1)})]^2/2}{2\tilde{\beta}pK+\tilde{\kappa}[pKt-pK(1-\xi_{\min}^{2(K-1)})]}\right) \tag{3.47}$$

$$\leq 2\exp\left(-\frac{\tilde{\kappa}^2 pK[t-(1-\xi_{\min}^{2(K-1)})]^2/2}{2\tilde{\beta}+\tilde{\kappa}[t-(1-\xi_{\min}^{2(K-1)})]}\right). \tag{3.48}$$

Further, when $t\in(1-\xi_{\min}^{2(K-1)}, 2-\xi_{\min}^{2(K-1)})$, we have $t-(1-\xi_{\min}^{2(K-1)})\leq 1$ and we arrive at

$$\mathbb{P}\left\{\left|\frac{1}{pK}\left\|\boldsymbol{O}^{(K)}\boldsymbol{x}\right\|^2-1\right|\geq t\right\}\leq 2\exp\left[-\tilde{c}pK(t+\xi_{\min}^{2(K-1)}-1)^2\right], \tag{3.49}$$

where $\tilde{c}=\frac{\tilde{\kappa}^2}{4\tilde{\beta}+2\tilde{\kappa}}>0$ depends only on β,κ.

Moreover, from Lemma 3.2, for $0<\rho<1/2$, we see that the restricted isometry constant satisfies

$$\mathbb{P}(\delta_s\geq\delta)\leq 2\left[\frac{en}{s}\left(1+\frac{2}{\rho}\right)\right]^s\exp\left[-\tilde{c}pK(\delta(1-2\rho)+\xi_{\min}^{2(K-1)}-1)^2\right], \tag{3.50}$$

when $\delta\in\left(\frac{1-\xi_{\min}^{2(K-1)}}{1-2\rho}, \min\left\{1, \frac{2-\xi_{\min}^{2(K-1)})}{1-2\rho}\right\}\right)$. However, we note that $2-\xi_{\min}^{2(K-1)}$ is an increasing function of K and

$$\frac{2-\xi_{\min}^{2(K-1)}}{1-2\rho}>\frac{1}{1-2\rho}>1. \tag{3.51}$$

So, $\min\left\{1, \frac{2-\xi_{\min}^{2(K-1)})}{1-2\rho}\right\}=1$. Consequently, for $\delta\in\left(\frac{1-\xi_{\min}^{2(K-1)}}{1-2\rho}, 1\right)$, it follows that $\mathbb{P}(\delta_s\leq\delta)$ with probability at least $1-\epsilon$ provided

$$\tilde{c}pK\left[\delta-\frac{1-\xi_{\min}^{2(K-1)}}{1-2\rho}\right]^2\geq\frac{1}{(1-2\rho)^2}\log(2\epsilon^{-1})+\frac{s}{(1-2\rho)^2}\log\left[\frac{en}{s}\left(1+\frac{2}{\rho}\right)\right]. \tag{3.52}$$

We choose $\rho=2/(e^7/2-1)<1/2$ so that

$$\frac{\log(1+2/\rho)}{(1-2\rho)^2}\leq\frac{14}{3}\quad\text{and}\quad\frac{1}{(1-2\rho)^2}\leq\frac{4}{3}. \tag{3.53}$$

So, $\mathbb{P}(\delta_s\leq\delta)$ for $\delta\in\left(\frac{2}{\sqrt{3}}(1-\xi_{\min}^{2(K-1)}), 1\right)$ with probability at least $1-\epsilon$ provided

$$pK\left[\delta-\frac{2}{\sqrt{3}}(1-\xi_{\min}^{2(K-1)})\right]^2\geq\frac{1}{\tilde{c}}\left[\frac{4}{3}\log(2\epsilon^{-1})+\frac{4}{3}\log(en/s)+\frac{14\,s}{3}\right] \tag{3.54}$$

$$= \frac{1}{\tilde{c}}\left[6s + \frac{4}{3}s\log(n/s) + \frac{4}{3}\log(2\epsilon^{-1})\right]. \quad (3.55)$$

Further bounding the measurement bound in the right-hand side above, we derive that $\mathbb{P}(\delta_s \leq \delta)$ with probability at least $1-\epsilon$ if

$$pK\left[\delta - \frac{2}{\sqrt{3}}(1-\xi_{\min}^{2(K-1)})\right]^2 \geq \frac{1}{\tilde{c}}\left[6s\log(en/s) + \frac{4}{3}\log(2\epsilon^{-1})\right]. \quad (3.56)$$

Hence, the proof is complete. □

If we relax Assumption 3.2 in Theorem 3.5, $\xi_{\min}$ is replaced with the ratio of the smallest to the largest singular value of $\boldsymbol{A}$.

Theorem 3.6 *Consider the linear dynamical system* ($\boldsymbol{A} \in \mathbb{R}^{n\times n}$, $\boldsymbol{B} \in \mathbb{R}^{n\times m}$, $\boldsymbol{C}_{(k)} \in \mathbb{R}^{p\times m}$, $k = 0, 1, \ldots, K-1$). *Let δ be such that*

$$c(1-\xi_r^{2(K-1)}) < \delta < 1, \quad (3.57)$$

where $0 < \xi_r < 1$ is the ratio of the smallest to the largest singular value of $\boldsymbol{A}$. Then, under Assumptions 3.1 and when $\boldsymbol{A} \neq \boldsymbol{0}$, there exist constants $c > \frac{2}{\sqrt{3}}$ and $C_1, C_2 > 0$ such that the restricted isometry constant of $\frac{1}{\sqrt{pK}}\boldsymbol{O}^{(K)}$ in (3.4) *satisfies $\delta_s \leq \delta$ with probability at least $1-\epsilon$ if, for $\epsilon \in (0,1)$,*

$$pK\left[\delta - c(1-\xi_r^{2(K-1)})\right]^2 \geq C_1 s\log(en/s) + C_2\log(2\epsilon^{-1}). \quad (3.58)$$

3.4 Sparse Observability

A natural consequence of Theorem 3.6 is the observability of a linear dynamical system with a sparse initial state.

Theorem 3.7 *Consider the linear dynamical system* ($\boldsymbol{A} \in \mathbb{R}^{n\times n}$, $\boldsymbol{B} \in \mathbb{R}^{n\times m}$, $\boldsymbol{C}_{(k)} \in \mathbb{R}^{p\times m}$, $k = 0, 1, \ldots, K-1$). *Then, under Assumption 3.1 and when $\boldsymbol{A} \neq \boldsymbol{0}$, there exist constants $c > \frac{2}{\sqrt{3}}$ and $C_1, C_2 > 0$ such that the linear system is observable for any s-sparse initial state with probability at least $1-\epsilon$ if, for any $\epsilon \in (0,1)$, some $K < \infty$ and $(\theta, \delta) \in \left\{(2, 1/3), (4, \sqrt{\sqrt{11/3}-1}/2), (3, 1/2)\right\}$*

$$\delta > c(1-\xi_r^{2(K-1)}) \quad (3.59)$$

$$pK\left[\delta - c(1-\xi_r^{2(K-1)})\right]^2 \geq C_1\theta s\log(en/\theta s) + C_2\log(2\epsilon^{-1}), \quad (3.60)$$

where $0 < \xi_r < 1$ is the ratio of the smallest to the largest singular value of $\boldsymbol{A}$.

Proof The result directly follows from Theorems 3.2, 3.3, 3.4 and 3.6. □

We next discuss some implications of the result.

3.4.1 Role of System Matrices

From Theorem 3.7, the observability depends on the number of rows p of the observation matrices and the ratio $0 < \xi_r < 1$ of the smallest to the largest singular value of $\boldsymbol{A}$.

Clearly, the linear system is observable with higher probability as the number of rows p of the observation matrices increases. Theorem 3.7 shows that $pK = \mathcal{O}(s\log(n/s))$ is sufficient for observability. In contrast, we need $pK = \mathcal{O}(n)$ for the classical non-sparse case. Therefore, for large dimensional systems, the number of measurements is much smaller than what is required by classical observability.

Furthermore, the best-case scenario for the ratio ξ_r is when $\xi_r = 1$. In that case, the result reduces to

$$pK\delta^2 \geq C_1\theta s\log(en/\theta s) + C_2\log(2\epsilon^{-1}). \tag{3.61}$$

As ξ_r moves away from one, the measurement bound increases, indicating that well-conditioned state matrices favor the observability of sparse initial states. Furthermore, when the state matrix is ill-conditioned, i.e., $\xi_r = 0$, then the system is sparse observable with high probability if

$$p\delta^2 \geq C_1\theta s\log(en/\theta s) + C_2\log(2\epsilon^{-1}). \tag{3.62}$$

This relation follows because when $\xi_r = 0$, under the above bound, we can recover any sparse initial state with $K = 1$.

In summary, there is a tradeoff between the number of measurements and the ratio ξ_r. We also note that the value of p for successful sparse recovery decreases with ξ_r and K. As K increases, we get more measurements, and a smaller p suffices for recovering the initial state. Also, as ξ_r increases, the matrix $\boldsymbol{A}$ becomes better conditioned, leading to a smaller value of p being sufficient.

3.4.2 Observability Index

The observability index refers to the smallest integer K^* such that $\boldsymbol{O}^{(K^*)} \in \mathbb{R}^{pK^* \times n}$ in (3.4) is full row rank, making the system observable. Mathematically,

$$K^* = \min_{k=1,2,\ldots} k \text{ s. t. } \operatorname{rank}\left\{\boldsymbol{O}^{(k)}\right\} = n. \tag{3.63}$$

The observability index is relatively easy to bound in the non-sparse case as follows:

Theorem 3.8 *The observability index K^* defined in* (3.63) *of an observable linear dynamical system* ($\boldsymbol{A} \in \mathbb{R}^{n\times n}, \boldsymbol{B} \in \mathbb{R}^{n\times m}, \boldsymbol{C}_{(k)} \in \mathbb{R}^{p\times m},\ k = 0, 1, \ldots, K-1$) *in* (3.1) *is bounded as*

$$\frac{n}{\max\limits_{k=0,1,\ldots,K^*-1} \text{rank}\left\{\boldsymbol{C}_{(k)}\right\}} \le K^* \le n - \text{rank}\left\{\boldsymbol{C}_{(0)}\right\} + 1. \tag{3.64}$$

When all the observation matrices have the same rank, i.e., $\text{rank}\left\{\boldsymbol{C}_{(k)}\right\} = \text{rank}\left\{\boldsymbol{C}\right\}$ *for* $k = 0, 1, \ldots, K^*$*, the bound reduces to*

$$\frac{n}{\text{rank}\left\{\boldsymbol{C}\right\}} \le K^* \le n - \text{rank}\left\{\boldsymbol{C}\right\} + 1. \tag{3.65}$$

Proof To prove the lower bound, we note from (3.4) that

$$n = \text{rank}\left\{\boldsymbol{O}^{(K^*)}\right\} \le \sum_{k=1}^{K^*} \text{rank}\left\{\boldsymbol{A}^{k-1}\boldsymbol{C}_{(k)}\right\} \tag{3.66}$$

$$\le \sum_{k=1}^{K^*} \text{rank}\left\{\boldsymbol{C}_{(k)}\right\} \tag{3.67}$$

$$= K^* \max_{k=0,1,\ldots,K^*-1} \text{rank}\left\{\boldsymbol{C}_{(k)}\right\}. \tag{3.68}$$

Rearranging the above relation, we arrive at the lower bound on K^*. Next, to prove the upper bound, we note that the number of linearly independent rows in $\boldsymbol{O}^{(1)} = \boldsymbol{C}_{(0)}$ is $\text{rank}\left\{\boldsymbol{C}_{(0)}\right\}$. Also, in the worst case, every block of p rows added to $\boldsymbol{O}^{(k)}$ for increasing values of k can only improve the rank by 1. So, $n \ge K^* - 1 + \text{rank}\left\{\boldsymbol{C}_{(0)}\right\}$ which implies that

$$K^* \le n - \text{rank}\left\{\boldsymbol{C}_{(0)}\right\} + 1. \tag{3.69}$$

Hence, the proof is complete. □

In the sparse initial state case, the bound changes as follows:

Theorem 3.9 *Let* K^* *be the observability index defined in* (3.63) *of an observable linear dynamical system* ($\boldsymbol{A} \in \mathbb{R}^{n\times n}$, $\boldsymbol{B} \in \mathbb{R}^{n\times m}$, $\boldsymbol{C}_{(k)} \in \mathbb{R}^{p\times m}$, $k = 0, 1, \ldots, K - 1$) *in* (3.1). *Under Assumption 3.1 and when* $\boldsymbol{A} \ne \boldsymbol{0}$*, there exist constants* $c > \frac{2}{\sqrt{3}}$ *and* $C_1, C_2 > 0$ *such that with probability at least* $1 - \epsilon$ *the observability index is bounded by*

$$\frac{1}{p\delta^2}[C_1\theta s \log(en/\theta s) + C_2 \log(2\epsilon^{-1})] \le K^* \le \left\lfloor \frac{\log 1/(1-\delta/c)}{2\log 1/\xi_r} \right\rfloor + 1 \tag{3.70}$$

where $(\theta, \delta) \in \left\{(2, 1/3), (4, \sqrt{\sqrt{11/3} - 1}/2), (3, 1/2)\right\}$.

Proof From Theorem 3.7, it follows that the system is observable for any s-sparse initial state with probability at least $1 - \epsilon$ if the following conditions holds:

$$\delta > c(1 - \xi_r^{2(K^*-1)}) \tag{3.71}$$

$$pK^* \left[\delta - c(1 - \xi_r^{2(K^*-1)})\right]^2 \geq C_1 \theta s \log(en/\theta s) + C_2 \log(2\epsilon^{-1}). \tag{3.72}$$

Simplifying (3.71), we derive that

$$K^* < \frac{\log 1/(1 - \delta/c)}{2 \log 1/\xi_r} + 1. \tag{3.73}$$

Considering that K^* is an integer, we arrive at the upper bound.

Also, using the fact that $\delta - c(1 - \xi_r^{2(K-1)})$ is a decreasing function of K, we get

$$\delta - c(1 - \xi_r^{2(K-1)}) \leq \delta, \tag{3.74}$$

which holds with equality when $K = 1$ or $\xi_r = 1$. By (3.72), we reach the relation,

$$K^* \geq \frac{1}{p\delta^2}[C_1 \theta s \log(en/\theta s) + C_2 \log(2\epsilon^{-1})]. \tag{3.75}$$

Thus, the proof is complete. □

In short, for the general (non-sparse) case, the observability index $K^* = \mathcal{O}(n)$. On the other hand, when the initial state is s-sparse, the observability index $K^* = \mathcal{O}(s \log(n/s))$.

3.5 Summary

- The concept of restricted isometry property of the observability matrix can be used to analyze the observability of a sparse initial state of a linear dynamical system.
- If the initial state of a linear dynamical system is sparse, the number of measurements or observations can be significantly reduced by using random observation matrices.
- The number of observations, given by the product of observability index and observation dimension, scales logarithmically with the state dimension and linearly with the sparsity level.

Notes

The predominant method for theoretically ensuring the exact recovery of an unknown sparse vector from underdetermined linear measurements is through the restricted isometry property. However, there are alternative metrics for evaluating the adequacy of the measurement matrix, such as coherence, which is not covered in our book. For a comprehensive exploration of these metrics, see Foucart and Rauhut (2013).

Various studies have indicated that a linear dynamical system achieves observability if the observability matrix fulfills the restricted isometry property Wakin et al. (2010); Dai and Yüksel (2013). Nonetheless, the conditions for the observability matrix to satisfy restricted isometry property are first introduced in Sanandaji et al. (2014). These findings, however, rely on stringent conditions like the system transfer matrix being unitary, the observation matrices being i.i.d. Gaussian, and the initial state being sparse on the canonical basis. Furthermore, these results depend on the smallest singular value of the state matrix, thus making them not invariant to the scaling of the state matrix. The examination of observability without constraints on the state matrix (as presented in this chapter) was detailed in Joseph and Murthy (2018). However, these outcomes are confined to scenarios where observation matrices are random and independent, and the initial state is sparse on the canonical basis. Subsequently, in Joseph and Murthy (2019), a probabilistic assurance for observability was derived under a sparse initial state assumption when the observation matrix remains fixed but is an instantiation of a random matrix. These results are also extended to the analysis of the settings where the initial state is sparse under a different basis other than the canonical basis.

References

Dai W, Yüksel S (2013) Observability of a linear system under sparsity constraints. IEEE Trans Autom Control 58(9):2372–2376

Foucart S, Rauhut H (2013) An invitation to compressive sensing. Springer

Joseph G, Murthy CR (2018) On the observability of a linear system with a sparse initial state. IEEE Signal Process Lett 25(7):994–998

Joseph G, Murthy CR (2019) Measurement bounds for observability of linear dynamical systems under sparsity constraints. IEEE Trans Signal Process 67(8):1992–2006

Sanandaji BM, Wakin MB, Vincent TL (2014) Observability with random observations. IEEE Trans Autom Control 59(11):3002–3007

Wakin MB, Sanandaji BM, Vincent TL (2010) On the observability of linear systems from random, compressive measurements. In: Proceedings of IEEE conference on decision and control, pp 4447–4454

Chapter 4
Sparse Controllability

Abstract In this chapter, we examine how the sparsity constraint on control inputs influences the controllability of linear dynamical systems. Specifically, we characterize the controllability of discrete-time linear dynamical systems with sparse control inputs. We derive algebraic necessary and sufficient conditions for ensuring controllability in systems with arbitrary transfer matrices. These conditions can be verified in polynomial time, in contrast to traditional Kalman-type rank tests. Additionally, we discuss the time-to-control of the system under these sparsity constraints on the inputs.

Keywords Kalman rank text · PBH test · Time-to-control · Sparse controllability index · Sparsity bounds · Sparse actuator schedule

4.1 Sparse Control

To define sparse control, we recall the state space model of the linear dynamical system ($\boldsymbol{A} \in \mathbb{R}^{n\times n}$, $\boldsymbol{B} \in \mathbb{R}^{n\times m}$), with state $\boldsymbol{x}_k \in \mathbb{R}^n$ and input $\boldsymbol{u}_k \in \mathbb{R}^m$,

$$\boldsymbol{x}_k = \boldsymbol{A}\boldsymbol{x}_{k-1} + \boldsymbol{B}\boldsymbol{u}_k, \tag{4.1}$$

for discrete-time indices $k = 1, 2, \ldots$. Here, we impose the constraint that the inputs are s-sparse for a given sparsity level, $\|\boldsymbol{u}_k\|_0 \leq s$, for $k = 1, 2, \ldots$. Also, we note that the support of the control inputs need not be identical for all k, i.e., we look at the time-varying support case.

An important aspect of the design of control systems is the ability to drive the system from any initial state to any desired state, called controllability. Formally, the linear dynamical system $(\boldsymbol{A}, \boldsymbol{B})$ in (4.1) is said to controllable if, for any initial state $\boldsymbol{x}_0$ and any final state $\boldsymbol{x}_{\mathrm{f}}$, there exist a time index $K < \infty$ and inputs $\{\boldsymbol{u}_k,\ k = 1, 2, \ldots, K\}$ that steer the system from the state $\boldsymbol{x}_0$ to $\boldsymbol{x}_K = \boldsymbol{x}_{\mathrm{f}}$. Controllability is critical in the analysis and control of systems as it ensures the feasibility of designing control inputs that can achieve the desired system performance.

G. Joseph and C. R. Murthy, *Sparsity-Constrained Linear Dynamical Systems*,
Springer Tracts in Electrical and Electronics Engineering,
https://doi.org/10.1007/978-981-97-7090-8_4

Our first goal in this chapter is to determine whether and how the new sparsity constraint changes the controllability of the system. To this end, we start with the classical notion of controllability and associated results.

4.2 Classical Controllability Theory

From the state space model in (4.1), we have

$$\boldsymbol{x}_K = \boldsymbol{A}^K \boldsymbol{x}_0 + \boldsymbol{\Phi}^{(K)} \boldsymbol{u}^{(K)}, \tag{4.2}$$

where $\boldsymbol{\Phi}^{(K)} \in \mathbb{R}^{n \times Km}$ is defined as

$$\boldsymbol{\Phi}^{(K)} = \left[(\boldsymbol{A}^{K-1}\boldsymbol{B})\ (\boldsymbol{A}^{K-2}\boldsymbol{B})\ \ldots\ \boldsymbol{B} \right], \tag{4.3}$$

and $\boldsymbol{u}^{(K)} \in \mathbb{R}^{Km}$ is formed by concatenating the K input vectors,

$$\boldsymbol{u}^{(K)} = \begin{bmatrix} \boldsymbol{u}_1 \\ \boldsymbol{u}_2 \\ \vdots \\ \boldsymbol{u}_K \end{bmatrix}. \tag{4.4}$$

The controllability of a linear dynamical system can be tested using the classical Kalman rank test or, equivalently, using the Popov–Belevitch–Hautus (PBH) test.

Theorem 4.1 *The linear dynamical system* ($\boldsymbol{A} \in \mathbb{R}^{n \times n}, \boldsymbol{B} \in \mathbb{R}^{n \times m}$) *in* (4.1) *is controllable if and only if it satisfies the Kalman rank condition given by*

$$\operatorname{rank} \left\{ \left[\boldsymbol{A}^{n-1}\boldsymbol{B}\ \boldsymbol{A}^{n-2}\boldsymbol{B}\ \ldots\ \boldsymbol{B} \right] \right\} = n. \tag{4.5}$$

Equivalently, the linear dynamical system ($\boldsymbol{A}, \boldsymbol{B}$) *in* (4.1) *is controllable if and only if it satisfies the PBH rank condition given by*

$$\operatorname{rank} \left\{ \left[\boldsymbol{A} - \lambda \boldsymbol{I}\ \boldsymbol{B} \right] \right\} = n, \quad \forall\, \lambda \in \mathbb{C}. \tag{4.6}$$

For controllable linear dynamical systems, we define the controllability index as follows. From the Kalman rank condition in (4.5), we infer that, for a controllable system ($\boldsymbol{A}, \boldsymbol{B}$), the controllability matrix $\boldsymbol{\Phi}^{(n)} \in \mathbb{R}^{n \times nm}$ in (4.3) has n linearly independent columns and the set of independent columns may not be unique. The controllability index refers to the smallest integer K^* such that $\boldsymbol{\Phi}^{(K^*)} \in \mathbb{R}^{n \times K^* m}$ has full row rank, i.e.,

$$K^* = \min_{k=1,2,\ldots} k \text{ s. t. } \operatorname{rank}\left\{ \boldsymbol{\Phi}^{(k)} \right\} = n. \tag{4.7}$$

Therefore, K^* denotes the least number of control inputs that can steer the system from any initial state to any desired final state at $k = K^*$. In other words, the controllability index, also called *time-to-control*, characterizes the minimum time required to steer the system from an arbitrary initial state to a desired state. The controllability index is bounded as follows:

Theorem 4.2 *The controllability index K^* defined in* (4.7) *of a controllable linear dynamical system* ($\boldsymbol{A} \in \mathbb{R}^{n\times n}$, $\boldsymbol{B} \in \mathbb{R}^{n\times m}$) *in* (4.1) *is bounded as*

$$\frac{n}{\operatorname{rank}\{\boldsymbol{B}\}} \leq K^* \leq n - \operatorname{rank}\{\boldsymbol{B}\} + 1. \tag{4.8}$$

Proof The proof is similar to that of Theorem 3.8. □

Next, we discuss the invariance of controllability under equivalent transformation.

Theorem 4.3 *The controllability property of a linear dynamical system* ($\boldsymbol{A} \in \mathbb{R}^{n\times n}$, $\boldsymbol{B} \in \mathbb{R}^{n\times m}$) *is invariant under reordering of* $\boldsymbol{B}$ *and equivalent transformation (i.e.,* ($\boldsymbol{P}^{-1}\boldsymbol{A}\boldsymbol{P}$, $\boldsymbol{P}^{-1}\boldsymbol{B}$) *for some nonsingular matrix* $\boldsymbol{P}$*).*

Proof Reordering of $\boldsymbol{B}$ leads to the reordering of columns of the controllability matrix $\boldsymbol{\Phi}^{(n)}$ in (4.3). Thus, the rank of $\boldsymbol{\Phi}^{(n)}$ does not change under reordering of $\boldsymbol{B}$, making the controllability property invariant under reordering of $\boldsymbol{B}$.

To prove the invariance under equivalent transformation, consider a similarity transformation of $\boldsymbol{A}$ given by $\boldsymbol{A} = \boldsymbol{P}\hat{\boldsymbol{A}}\boldsymbol{P}^{-1}$, where $\boldsymbol{P}$ is a nonsingular matrix. Then, with $\hat{\boldsymbol{B}} = \boldsymbol{P}^{-1}\boldsymbol{B}$, we have

$$\operatorname{rank}\left\{\boldsymbol{\Phi}^{(n)}\right\} = \operatorname{rank}\left\{\left[\boldsymbol{A}^{n-1}\boldsymbol{B}\ \boldsymbol{A}^{n-2}\boldsymbol{B}\ \ldots\ \boldsymbol{B}\right]\right\} \tag{4.9}$$

$$= \operatorname{rank}\left\{\boldsymbol{P}\left[\hat{\boldsymbol{A}}^{n-1}\hat{\boldsymbol{B}}\ \hat{\boldsymbol{A}}^{n-2}\hat{\boldsymbol{B}}\ \ldots\ \hat{\boldsymbol{B}}\right]\right\} \tag{4.10}$$

$$= \operatorname{rank}\left\{\left[\hat{\boldsymbol{A}}^{n-1}\hat{\boldsymbol{B}}\ \hat{\boldsymbol{A}}^{n-2}\hat{\boldsymbol{B}}\ \ldots\ \hat{\boldsymbol{B}}\right]\right\}, \tag{4.11}$$

which is the rank of the controllability matrix of the linear dynamical system ($\boldsymbol{P}^{-1}\boldsymbol{A}\boldsymbol{P}$, $\boldsymbol{P}^{-1}\boldsymbol{B}$). Hence, the controllability property is invariant to equivalent transformation. □

In summary, this section briefly outlined the important controllability-related results from the classical control theory. We next discuss how the above results change when we introduce the sparsity constraints described in Sect. 4.1.

4.3 Sparse Controllability

We define a modified version of controllability under sparsity constraints as follows.

Definition 4.1 The linear dynamical system ($\boldsymbol{A} \in \mathbb{R}^{n\times n}$, $\boldsymbol{B} \in \mathbb{R}^{n\times m}$) in (4.1) is said to be s-sparse controllable for a given $s \geq 1$ if, for any initial state $\boldsymbol{x}_0 \in \mathbb{R}^n$ and any final state $\boldsymbol{x}_\mathrm{f} \in \mathbb{R}^n$, there exist a time index $K < \infty$ and inputs $\{\boldsymbol{u}_k \in \mathbb{R}^m : \|\boldsymbol{u}_k\|_0 \leq s,\ k = 1, 2, \ldots, K\}$ that steer the system from the state $\boldsymbol{x}_0$ to the state $\boldsymbol{x}_K = \boldsymbol{x}_\mathrm{f}$.

We first derive a controllability test similar to the Kalman rank test for sparse controllability.

Theorem 4.4 *The linear dynamical system* ($\boldsymbol{A} \in \mathbb{R}^{n\times n}$, $\boldsymbol{B} \in \mathbb{R}^{n\times m}$) *in* (4.1) *is* s-*sparse controllable if and only if there exist an integer* $K < \infty$ *and sets* $\{\mathcal{S}_k \subseteq \{1, 2, \ldots, m\},\ |\mathcal{S}_k| \leq s\}_{k=1}^K$ *such that*

$$\operatorname{rank}\left\{\left[\boldsymbol{A}^{K-1}\boldsymbol{B}_{\mathcal{S}_1}\ \boldsymbol{A}^{K-2}\boldsymbol{B}_{\mathcal{S}_2}\ \ldots\ \boldsymbol{B}_{\mathcal{S}_K}\right]\right\} = n, \tag{4.12}$$

where $\boldsymbol{B}_{\mathcal{S}_k} \in \mathbb{R}^{n\times|\mathcal{S}_k|}$ *is the submatrix formed by the columns of* $\boldsymbol{B}$ *indexed by* $\mathcal{S}_k$, *for any* $k = 1, 2, \ldots, K$.

Proof From Definition 4.1, we see that the system is s-sparse controllable if and only if, for any vectors $\boldsymbol{x}_0$ and $\boldsymbol{x}_\mathrm{f}$, there exist an integer $K < \infty$ and $\{\boldsymbol{u}_k \in \mathbb{R}^m : \|\boldsymbol{u}_k\|_0 \leq s\}_{k=1}^K$ such that

$$\boldsymbol{x}_K - \boldsymbol{A}^K\boldsymbol{x}_0 = \left[\boldsymbol{A}^{K-1}\boldsymbol{B}\ \boldsymbol{A}^{K-2}\boldsymbol{B}\ \ldots\ \boldsymbol{B}\right]\begin{bmatrix}\boldsymbol{u}_1\\ \boldsymbol{u}_2\\ \vdots\\ \boldsymbol{u}_K\end{bmatrix}. \tag{4.13}$$

So, the system is s-sparse controllable if and only if, for any vectors $\boldsymbol{x}_0$ and $\boldsymbol{x}_\mathrm{f}$, there exist an integer $K < \infty$, support sets $\{\mathcal{S}_k \subseteq \{1, 2, \ldots, m\},\ |\mathcal{S}_k| \leq s\}_{k=1}^K$, and $\tilde{\boldsymbol{u}} \in \mathbb{R}^{\sum_{k=1}^K |\mathcal{S}_k|}$ such that

$$\boldsymbol{x}_K - \boldsymbol{A}^K\boldsymbol{x}_0 = \left[\boldsymbol{A}^{K-1}\boldsymbol{B}_{\mathcal{S}_1}\ \boldsymbol{A}^{K-2}\boldsymbol{B}_{\mathcal{S}_2}\ \ldots\ \boldsymbol{B}_{\mathcal{S}_K}\right]\tilde{\boldsymbol{u}}. \tag{4.14}$$

Hence, the system is s-sparse controllable if and only if

$$\mathbb{R}^n = \bigcup_{\substack{\{\mathcal{S}_k\subseteq\{1,2,\ldots,m\},\\ |\mathcal{S}_k|\leq s\}_{k=1}^K}} \operatorname{ColSpace}\left\{\left[\boldsymbol{A}^{K-1}\boldsymbol{B}_{\mathcal{S}_1}\ \boldsymbol{A}^{K-2}\boldsymbol{B}_{\mathcal{S}_2}\ \ldots\ \boldsymbol{B}_{\mathcal{S}_K}\right]\right\}, \tag{4.15}$$

where ColSpace$\{\cdot\}$ denotes the column space of a matrix. Further, the above relation implies that the union of a finite number of vector spaces is another vector space, which happens if and only if one of the vector spaces contains every other vector space. As a result, Statement 4.5 holds if and only if there exist an integer $K < \infty$ and support sets $\{\mathcal{S}_k \subseteq \{1, 2, \ldots, m\},\ |\mathcal{S}_k| \leq s\}_{k=1}^{K}$, such that

$$\mathbb{R}^n = \text{ColSpace}\left\{\left[\boldsymbol{A}^{K-1}\boldsymbol{B}_{\mathcal{S}_1}\ \boldsymbol{A}^{K-2}\boldsymbol{B}_{\mathcal{S}_2}\ \ldots\ \boldsymbol{B}_{\mathcal{S}_K}\right]\right\}, \tag{4.16}$$

which is equivalent to the desired sparse controllability test. Thus, the proof is complete. □

The above result gives an interesting insight into sparse control with time-varying support. By definition, the support set at any time index can vary depending on the initial and final states. However, Theorem 4.4 shows that, if the system is sparse controllable, there exists a common actuator schedule that works for any initial and final states.

Further, Theorem 4.4 implies the invariance of sparse controllability under equivalent transformation.

Proposition 4.1 *The s-sparse controllability property of a linear dynamical system ($\boldsymbol{A} \in \mathbb{R}^{n\times n}$, $\boldsymbol{B} \in \mathbb{R}^{n\times m}$) is invariant under reordering of $\boldsymbol{B}$ and equivalent transformation.*

Proof The proof is similar to that of Theorem 4.3. □

Also, Theorem 4.4 provides insights into sparse controllability for some special cases.

Corollary 4.1 *A controllable linear dynamical system ($\boldsymbol{A} \in \mathbb{R}^{n\times n}$, $\boldsymbol{B} \in \mathbb{R}^{n\times m}$) is s-sparse controllable if $s >$ rank $\{\boldsymbol{B}\}$. In particular, when* rank $\{\boldsymbol{B}\} = 1$, *sparse controllability is equivalent to controllability.*

Proof If $s >$ rank $\{\boldsymbol{B}\}$, then we can find s columns indexed by $\mathcal{S}$ of $\boldsymbol{B}$ that spans its column space. Then, using Theorem 4.1, we have

$$n = \text{rank}\left\{\left[\boldsymbol{A}^{K-1}\boldsymbol{B}\ \boldsymbol{A}^{K-2}\boldsymbol{B}\ \ldots\ \boldsymbol{B}\right]\right\} \tag{4.17}$$

$$= \text{rank}\left\{\left[\boldsymbol{A}^{K-1}\boldsymbol{B}_{\mathcal{S}}\ \boldsymbol{A}^{K-2}\boldsymbol{B}_{\mathcal{S}}\ \ldots\ \boldsymbol{B}_{\mathcal{S}}\right]\right\}. \tag{4.18}$$

Then, by Theorem 4.4, the system is s-sparse controllable. □

Although Theorem 4.4 gives a sparse controllability test, this condition can not be verified easily because the number of possible support sets is exponential in n. Moreover, similar to the Kalman rank test, the above test also requires us to compute different powers of $\boldsymbol{A}$, which can lead to numerical instability. So, we next present a simpler condition to test sparse controllability.

Theorem 4.5 *For the linear dynamical system ($\boldsymbol{A} \in \mathbb{R}^{n\times n}$, $\boldsymbol{B} \in \mathbb{R}^{n\times m}$) in (4.1), the following statements are equivalent:*

(a) The system ($\boldsymbol{A}$, $\boldsymbol{B}$) is s-sparse controllable (see Definition 4.1).

(b) *There exist an integer* $K < \infty$ *and sets* $\{\mathcal{S}_k \subseteq \{1, 2, \ldots, m\},\ |\mathcal{S}_k| \le s\}_{k=1}^{K}$ *such that*

$$\operatorname{rank}\left\{\left[\boldsymbol{A}^{K-1}\boldsymbol{B}_{\mathcal{S}_1}\ \boldsymbol{A}^{K-2}\boldsymbol{B}_{\mathcal{S}_2}\ \ldots\ \boldsymbol{B}_{\mathcal{S}_K}\right]\right\} = n. \tag{4.19}$$

(c) *For any nonzero* $\lambda \in \mathbb{C}$, *we have* $\operatorname{rank}\left\{\left[\boldsymbol{A} - \lambda\boldsymbol{I}\ \boldsymbol{B}\right]\right\} = n$ *and there exists a set* $\mathcal{S} \subseteq \{1, 2, \ldots, m\}$ *such that* $1 \le |\mathcal{S}| \le s$ *and* $\operatorname{rank}\left\{\left[\boldsymbol{A}\ \boldsymbol{B}_{\mathcal{S}}\right]\right\} = n$.

Proof The equivalence of Statements (a) and (b) was shown in Theorem 4.4. Therefore, it suffices to establish the equivalence of Statements (b) and (c). We first show that Statement (b) implies Statement (c). Suppose that Statement (b) holds. We derive that

$$\mathbb{R}^n = \operatorname{ColSpace}\left\{\left[\boldsymbol{A}^{K-1}\boldsymbol{B}_{\mathcal{S}_1}\ \boldsymbol{A}^{K-2}\boldsymbol{B}_{\mathcal{S}_2}\ \ldots\ \boldsymbol{B}_{\mathcal{S}_K}\right]\right\} \tag{4.20}$$

$$\subseteq \operatorname{ColSpace}\left\{\left[\boldsymbol{A}^{K-1}\boldsymbol{B}\ \boldsymbol{A}^{K-2}\boldsymbol{B}\ \ldots\ \boldsymbol{B}\right]\right\} \tag{4.21}$$

$$\subseteq \mathbb{R}^n. \tag{4.22}$$

Hence, we have

$$\operatorname{rank}\left\{\left[\boldsymbol{A}^{K-1}\boldsymbol{B}\ \boldsymbol{A}^{K-2}\boldsymbol{B}\ \ldots\ \boldsymbol{B}\right]\right\} = n. \tag{4.23}$$

Now, using the equivalence of the Kalman and PHB rank conditions by Theorem 4.1, we conclude that, for all $\lambda \in \mathbb{C}$,

$$\operatorname{rank}\left\{\left[\boldsymbol{A} - \lambda\boldsymbol{I}\ \boldsymbol{B}\right]\right\} = n, \tag{4.24}$$

which establishes the first condition of Statement (c). Further, we note that, if Statement (b) holds,

$$\mathbb{R}^n = \operatorname{ColSpace}\left\{\left[\boldsymbol{A}^{K-1}\boldsymbol{B}_{\mathcal{S}_1}\ \boldsymbol{A}^{K-2}\boldsymbol{B}_{\mathcal{S}_2}\ \ldots\ \boldsymbol{B}_{\mathcal{S}_K}\right]\right\} \tag{4.25}$$

$$\subseteq \operatorname{ColSpace}\left\{\left[\boldsymbol{A}\ \boldsymbol{B}_{\mathcal{S}_K}\right]\right\} \subseteq \mathbb{R}^n, \tag{4.26}$$

which follows from the fact that the first $\sum_{k=1}^{K-1} |\mathcal{S}_k|$ columns of the matrix belong to the column space of $\boldsymbol{A}$. Hence, Statement (b) implies Statement (c).

Next, to prove the equivalence between Statements (b) and (c), we show that, if Statement (b) does not hold, Statement (c) is also violated. Suppose that, for any $K < \infty$,

$$\operatorname{rank}\left\{\left[\boldsymbol{A}^{K-1}\boldsymbol{B}_{\mathcal{S}_1}\ \boldsymbol{A}^{K-2}\boldsymbol{B}_{\mathcal{S}_2}\ \ldots\ \boldsymbol{B}_{\mathcal{S}_K}\right]\right\} < n, \tag{4.27}$$

for any sets $\{\mathcal{S}_k \subseteq \{1, 2, \ldots, m\},\ |\mathcal{S}_k| \le s\}_{k=1}^{K}$. Let $\mathcal{S} \subseteq \{1, 2, \ldots, m\}$ such that $1 \le |\mathcal{S}| \le s$. We can construct sets $\mathcal{T}_k \subseteq \{1, 2, \ldots, m\}$ for $k = 1, 2, \ldots, m-1$ such that $|\mathcal{T}_k| \le s$ and

$$\bigcup_{k=1}^{m-1} \mathcal{T}_k \bigcup \mathcal{S} = \{1, 2, \ldots, m\}. \tag{4.28}$$

Then, using the assumption in (4.27), with $K = nm$, we get

$$\begin{aligned}\text{rank}\{ [\, & \boldsymbol{A}^{nm-1}\boldsymbol{B}_{\mathcal{T}_1}\ \boldsymbol{A}^{nm-2}\boldsymbol{B}_{\mathcal{T}_1}\ \ldots\ \boldsymbol{A}^{n(m-1)}\boldsymbol{B}_{\mathcal{T}_1} \\ & \boldsymbol{A}^{n(m-1)-1}\boldsymbol{B}_{\mathcal{T}_2}\ \boldsymbol{A}^{n(m-1)-2}\boldsymbol{B}_{\mathcal{T}_2}\ \ldots\ \boldsymbol{A}^{n(m-2)}\boldsymbol{B}_{\mathcal{T}_2} \\ & \ldots\ \boldsymbol{A}^{2n-1}\boldsymbol{B}_{\mathcal{T}_{m-1}}\ \boldsymbol{A}^{2n-2}\boldsymbol{B}_{\mathcal{T}_{m-1}}\ \ldots\ \boldsymbol{A}^{n}\boldsymbol{B}_{\mathcal{T}_{m-1}} \\ & \ldots\ \boldsymbol{A}^{n-1}\boldsymbol{B}_{\mathcal{S}}\ \boldsymbol{A}^{n-2}\boldsymbol{B}_{\mathcal{S}}\ \ldots\ \boldsymbol{B}_{\mathcal{S}}\,]\} < n. \end{aligned} \tag{4.29}$$

We rearrange the columns of the above matrix to get

$$\begin{aligned}\text{rank}\{[\, & \boldsymbol{A}^{nm-1}\boldsymbol{B}_{\mathcal{T}_1}\ \boldsymbol{A}^{n(m-1)-1}\boldsymbol{B}_{\mathcal{T}_2}\ \ldots\ \boldsymbol{A}^{2n-1}\boldsymbol{B}_{\mathcal{T}_{m-1}}\ \boldsymbol{A}^{n-1}\boldsymbol{B}_{\mathcal{S}} \\ & \boldsymbol{A}^{nm-2}\boldsymbol{B}_{\mathcal{T}_1}\ \boldsymbol{A}^{n(m-1)-2}\boldsymbol{B}_{\mathcal{T}_2}\ \ldots\ \boldsymbol{A}^{2n-2}\boldsymbol{B}_{\mathcal{T}_{m-1}}\ \boldsymbol{A}^{n-2}\boldsymbol{B}_{\mathcal{S}} \\ & \ldots \boldsymbol{A}^{n(m-1)}\boldsymbol{B}_{\mathcal{T}_1}\ \boldsymbol{A}^{n(m-2)}\boldsymbol{B}_{\mathcal{T}_2} \ldots\ \boldsymbol{A}^{n}\boldsymbol{B}_{\mathcal{T}_{m-1}}\ \boldsymbol{B}_{\mathcal{S}}\,]\} < n. \end{aligned} \tag{4.30}$$

Now we define the following matrix,

$$\hat{\boldsymbol{B}} = \left[\boldsymbol{A}^{n(m-1)}\boldsymbol{B}_{\mathcal{T}_1}\ \boldsymbol{A}^{n(m-2)}\boldsymbol{B}_{\mathcal{T}_2} \ldots\ \boldsymbol{A}^{n}\boldsymbol{B}_{\mathcal{T}_{m-1}}\ \boldsymbol{B}_{\mathcal{S}}\right], \tag{4.31}$$

to rewrite (4.30) as

$$\text{rank}\left\{\left[\boldsymbol{A}^{n-1}\hat{\boldsymbol{B}}\ \boldsymbol{A}^{n-2}\hat{\boldsymbol{B}}\ \ldots\ \hat{\boldsymbol{B}}\right]\right\} < n. \tag{4.32}$$

Next, from the Kalman rank condition in Theorem 4.1, we conclude that the system $(\boldsymbol{A}, \hat{\boldsymbol{B}})$ is not controllable and it violates the PBH rank condition. So there exists $\lambda \in \mathbb{C}$ such that

$$\text{rank}\left\{\left[\boldsymbol{A} - \lambda \boldsymbol{I}\ \hat{\boldsymbol{B}}\right]\right\} < n. \tag{4.33}$$

Consequently, there exists $z \in \mathbb{C}^n$ such that $z \neq \boldsymbol{0}$ and

$$z^{\mathsf{T}}\left[\boldsymbol{A} - \lambda \boldsymbol{I}\ \hat{\boldsymbol{B}}\right] = \boldsymbol{0}. \tag{4.34}$$

This relation leads us to

$$z^{\mathsf{T}}\boldsymbol{A} = \lambda z^{\mathsf{T}} \tag{4.35}$$

$$z^{\mathsf{T}}\hat{\boldsymbol{B}} = \boldsymbol{0}. \tag{4.36}$$

Substituting for $\hat{\boldsymbol{B}}$ from (4.31) and using (4.35), we arrive at

$$\boldsymbol{0} = z^{\mathsf{T}}\hat{\boldsymbol{B}} = z^{\mathsf{T}}\left[\boldsymbol{A}^{n(m-1)}\boldsymbol{B}_{\mathcal{T}_1}\ \boldsymbol{A}^{n(m-2)}\boldsymbol{B}_{\mathcal{T}_2} \ldots\ \boldsymbol{A}^{n}\boldsymbol{B}_{\mathcal{T}_{m-1}}\ \boldsymbol{B}_{\mathcal{S}}\right] \tag{4.37}$$

$$= z^{\mathsf{T}}\left[\lambda^{n(m-1)}\boldsymbol{B}_{\mathcal{T}_1}\ \lambda^{n(m-2)}\boldsymbol{B}_{\mathcal{T}_2} \ldots\ \lambda^{n}\boldsymbol{B}_{\mathcal{T}_{m-1}}\ \boldsymbol{B}_{\mathcal{S}}\right]. \tag{4.38}$$

We now consider two cases: $\lambda = 0$ and $\lambda \neq 0$ separately as follows.

If $\lambda = 0$, from (4.38), we derive

$$\mathbf{0} = z^{\mathsf{T}} \boldsymbol{B}_{\mathcal{S}}. \tag{4.39}$$

This condition combined with (4.35) implies

$$\mathbf{0} = z^{\mathsf{T}} \left[\boldsymbol{A} - \lambda \boldsymbol{I} \; \boldsymbol{B}_{\mathcal{S}}\right] = z^{\mathsf{T}} \left[\boldsymbol{A} \; \boldsymbol{B}_{\mathcal{S}}\right]. \tag{4.40}$$

Hence, for any set $\mathcal{S} \subseteq \{1, 2, \ldots, m\}$, we derive $\operatorname{rank}\left\{\left[\boldsymbol{A} \; \boldsymbol{B}_{\mathcal{S}}\right]\right\} < n$, violating Statement (c).

If $\lambda \neq 0$, from (4.38), we derive

$$\mathbf{0} = z^{\mathsf{T}} \left[\boldsymbol{B}_{\mathcal{T}_1} \; \boldsymbol{B}_{\mathcal{T}_2} \ldots \boldsymbol{B}_{\mathcal{T}_{m-1}} \; \boldsymbol{B}_{\mathcal{S}}\right] = z^{\mathsf{T}} \boldsymbol{B}, \tag{4.41}$$

using (4.28). This condition combined with (4.35) implies

$$\mathbf{0} = z^{\mathsf{T}} \left[\boldsymbol{A} - \lambda \boldsymbol{I} \; \boldsymbol{B}\right]. \tag{4.42}$$

Hence, there exists $\lambda \neq 0$ such that $\operatorname{rank}\left\{\left[\boldsymbol{A} - \lambda \boldsymbol{I} \; \boldsymbol{B}\right]\right\} < n$, violating Statement (c).

To summarize, Statement (c) is violated if Statement (b) does not hold. So Statement (c) implies Statement (b), and the proof is complete. □

The proof technique of Theorem 4.5 is quite general and can be extended to other sparsity structures.

Corollary 4.2 *The linear dynamical system* ($\boldsymbol{A} \in \mathbb{R}^{n\times n}, \boldsymbol{B} \in \mathbb{R}^{n\times m}$) *in* (4.1) *with feasible inputs satisfying the constraint* $\operatorname{Supp}(\boldsymbol{u}_k) \in \mathbb{S}$, *where* $\mathbb{S}$ *is a subset of the powerset of* $\{1, 2, \ldots, m\}$ *satisfying* $\bigcup_{\mathcal{S}\in\mathbb{S}} = \{1, 2, \ldots, m\}$. *Then, the system is controllable under the above constraint if and only if, for any nonzero* $\lambda \in \mathbb{C}$, *we have* $\operatorname{rank}\left\{\left[\boldsymbol{A} - \lambda \boldsymbol{I} \; \boldsymbol{B}\right]\right\} = n$ *and there exists a set* $\mathcal{S} \in \mathbb{S}$ *such that* $\operatorname{rank}\left\{\left[\boldsymbol{A} \; \boldsymbol{B}_{\mathcal{S}}\right]\right\} = n$.

Proof The proof is identical to the proof of Theorem 4.5 except that instead of choosing support sets with cardinality at most s, we choose the support sets from $\mathbb{S}$. □

We note that Theorem 4.5 is a special case of Corollary 4.2 where $\mathbb{S} = \{\mathcal{S} \subseteq \{1, 2, \ldots, m\} : |\mathcal{S}| \leq s\}$. Further, the conditions of the new controllability test, Statement (c) of Theorem 4.5 is simpler than that given by Theorem 4.4. Nonetheless, the second condition needs to be verified for support sets of size s, which might appear to be a combinatorial problem. However, the condition can be further simplified as follows.

Theorem 4.6 *The linear dynamical system* ($\boldsymbol{A} \in \mathbb{R}^{n\times n}, \boldsymbol{B} \in \mathbb{R}^{n\times m}$) *in* (4.1) *is* s-*sparse controllable if and only if the system is controllable and the sparsity level* $s \geq \max\{1, n - \operatorname{rank}\{\boldsymbol{A}\}\}$.

Proof From Theorem 4.1 and Theorem 4.5, it suffices to establish that the following statements are equivalent:

1. For any nonzero $\lambda \in \mathbb{C}$, we have $\operatorname{rank}\left\{\left[\boldsymbol{A} - \lambda \boldsymbol{I} \ \boldsymbol{B}\right]\right\} = n$ and there exists a set $\mathcal{S}$ such that $1 \leq |\mathcal{S}| \leq s$ and $\operatorname{rank}\left\{\left[\boldsymbol{A} \ \boldsymbol{B}_{\mathcal{S}}\right]\right\} = n$.
2. For any $\lambda \in \mathbb{C}$, we have $\operatorname{rank}\left\{\left[\boldsymbol{A} - \lambda \boldsymbol{I} \ \boldsymbol{B}\right]\right\} = n$ and $s \geq \max\{1, n - \operatorname{rank}\{\boldsymbol{A}\}\}$.

We first show that Statement 1 implies Statement 2. Statement 1 immediately implies the first condition of Statement 2. Also, if Statement 1 holds, we derive

$$n = \operatorname{rank}\left\{\left[\boldsymbol{A} \ \boldsymbol{B}_{\mathcal{S}}\right]\right\} \leq \operatorname{rank}\{\boldsymbol{A}\} + \operatorname{rank}\{\boldsymbol{B}_{\mathcal{S}}\} \leq \operatorname{rank}\{\boldsymbol{A}\} + s. \tag{4.43}$$

As a consequence, we deduce that $s \geq n - \operatorname{rank}\{\boldsymbol{A}\}$. Hence, we establish the second condition of Statement 2, and prove that Statement 1 implies Statement 2.

Next, we show that Statement 2 implies Statement 1. Statement 2 immediately implies the first condition of Statement 1 and $\operatorname{rank}\left\{\left[\boldsymbol{A} \ \boldsymbol{B}\right]\right\} = n$. Then, there exist $n - \operatorname{rank}\{\boldsymbol{A}\}$ columns in $\boldsymbol{B}$ that are linearly independent of the columns of $\boldsymbol{A}$. So if $s \geq n - \operatorname{rank}\{\boldsymbol{A}\}$, we can always find a support set $\mathcal{S}$ with $|\mathcal{S}| = n - \operatorname{rank}\{\boldsymbol{A}\} \leq s$ such that $\operatorname{rank}\left\{\left[\boldsymbol{A} \ \boldsymbol{B}_{\mathcal{S}}\right]\right\} = n$. Hence, Statement 2 implies Statement 1, and the proof is complete. □

The first condition of Theorem 4.6 is the same as the PBH test, which is not surprising. If a system is not controllable (without any restrictions on the input), it is not sparse controllable. So, the PBH test is a necessary condition for sparse controllability. Theorem 4.6 establishes that any controllable system is sparse controllable if the lower limit on sparsity is met. Also, the first condition is independent of the sparsity level, and the lower limit in the second condition does not depend on the input matrix $\boldsymbol{B}$. Furthermore, the PBH test also provides the rank of $\boldsymbol{A}$ and therefore, the controllability test in Theorem 4.6 has the same complexity as the classical PBH test in Theorem 4.1. So, unlike the combinatorial test in Theorem 4.4, the new controllability test has polynomial complexity in n, and the complexity is independent of the sparsity level s.

Another important insight from Theorem 4.6 about reversible linear dynamical systems i.e., linear dynamical systems whose state matrix $\boldsymbol{A}$ is invertible, is as follows:

Corollary 4.3 *A controllable linear dynamical system* $(\boldsymbol{A} \in \mathbb{R}^{n\times n}, \boldsymbol{B} \in \mathbb{R}^{n\times m})$ *in* (4.1) *with* $\operatorname{rank}\{\boldsymbol{A}\} \geq n - 1$ *is s-sparse controllable for any* $s \geq 1$.

So far, we have discussed the conditions under which a linear dynamical system is sparse controllable. Next, we discuss the controllability index under sparsity constraints.

4.4 Sparse Controllability Index

The controllability index K_s^* under the constraint that the inputs are s-sparse can be defined similarly to the unconstrained case.

$$K_s^* = \min_{\substack{\{\mathcal{S}_k \subseteq \{1,2,\ldots,m\},\\ |\mathcal{S}_k| \leq s\}_{k=1}^K}} K \text{s. t. } \operatorname{rank}\left\{\left[A^{K-1} B_{\mathcal{S}_1}\ A^{K-2} B_{\mathcal{S}_2} \ \ldots\ B_{\mathcal{S}_K}\right]\right\} = n. \tag{4.44}$$

Therefore, K_s^* denotes the time-to-control or the least number of control inputs that can steer the system from any initial state to any desired final state at $k = K_s^*$, using s-sparse inputs.

We next bound the sparse controllability index under sparsity constraints, similar to Theorem 4.2. We start with the lower bound.

Lemma 4.1 *The s-sparse controllability index K_s^* defined in* (4.44) *of a controllable linear dynamical system* ($A \in \mathbb{R}^{n\times n}$, $B \in \mathbb{R}^{n\times m}$) *in* (4.1) *is bounded as*

$$K_s^* \geq \frac{n}{\min\{\operatorname{rank}\{B\}, s\}}. \tag{4.45}$$

Proof To prove the lower bound, let $\{\mathcal{S}_k \subseteq \{1, 2, \ldots, m\},\ |\mathcal{S}_k| \leq s\}_{k=1}^{K_s^*}$ such that

$$n = \operatorname{rank}\left\{\left[A^{K_s^*-1} B_{\mathcal{S}_1}\ A^{K-2} B_{\mathcal{S}_2} \ \ldots\ B_{\mathcal{S}_{K_s^*}}\right]\right\}. \tag{4.46}$$

Then, we deduce that

$$n \leq \sum_{k=1}^{K_s^*} \operatorname{rank}\left\{A^{K_s^*-k} B_{\mathcal{S}_k}\right\} \leq \sum_{k=1}^{K^*} \operatorname{rank}\left\{B_{\mathcal{S}_k}\right\} = K^* \min\{\operatorname{rank}\{B\}, s\}. \tag{4.47}$$

Rearranging the above relation, we arrive at the lower bound on K_s^*. □

We next derive the upper bound for the sparse controllability index using the following lemma.

Lemma 4.2 *For a controllable linear dynamical system* ($A \in \mathbb{R}^{n\times n}$, $B \in \mathbb{R}^{n\times m}$) *in* (4.1)*, we define* $\Phi^{(K,s)}$ *as*

$$\begin{aligned}
\Phi^{(K,s)} \in \ & \operatorname*{arg\,max}_{\Phi \in \mathbb{R}^{n\times Ks}} \operatorname{rank}\{\Phi\} \\
& \text{s. t. } \Phi = \left[A^{K-1} B_{\mathcal{S}_1}\ A^{K-2} B_{\mathcal{S}_2} \ \ldots\ B_{\mathcal{S}_K}\right], \\
& \text{and } \mathcal{S}_k \subseteq \{1, 2, \ldots, m\},\ \ |\mathcal{S}_k| \leq s,\ \ k = 1, 2, \ldots, K.
\end{aligned} \tag{4.48}$$

Then, for any $K \leq K_s^*$, the rank of the matrix satisfies

$$\operatorname{rank}\left\{\Phi^{(K,s)}\right\} > \operatorname{rank}\left\{\Phi^{(K-1,s)}\right\}. \tag{4.49}$$

Proof We notice that $K_s^* = \min\{K : \operatorname{rank}\left\{\Phi^{(K,s)}\right\} = n\}$ and

$$\operatorname{rank}\left\{\Phi^{(K_s^*,s)}\right\} = n > \operatorname{rank}\left\{\Phi^{(K_s^*-1,s)}\right\}, \tag{4.50}$$

establishing the desired result for $K = K_s^*$.

Next, we prove the result for $K < K_s^*$. Suppose (4.49) is not true for some $K < K_s^*$, i.e., $\text{rank}\left\{\boldsymbol{\Phi}^{(K,s)}\right\} \leq \text{rank}\left\{\boldsymbol{\Phi}^{(K-1,s)}\right\}$, and let $\mathcal{T}_k \subseteq \{1, 2, \ldots, m\}$ with $|\mathcal{T}_k| \leq s$, for $k = 1, 2, \ldots, K-1$, be such that

$$\text{rank}\left\{\boldsymbol{\Phi}^{(K-1,s)}\right\} = \text{rank}\left\{\left[\boldsymbol{A}^{K-2}\boldsymbol{B}_{\mathcal{T}_1}\ \boldsymbol{A}^{K-3}\boldsymbol{B}_{\mathcal{T}_2}\ \ldots\ \boldsymbol{B}_{\mathcal{T}_{K-1}}\right]\right\}. \tag{4.51}$$

Then, for any $\mathcal{S} \subseteq \{1, 2, \ldots, m\}$ with $|\mathcal{S}| \leq s$, we have

$$\begin{aligned}\text{rank}\left\{\boldsymbol{\Phi}^{(K,s)}\right\} &\geq \text{rank}\left\{\left[\boldsymbol{A}^{K-1}\boldsymbol{B}_{\mathcal{S}}\ \boldsymbol{A}^{K-2}\boldsymbol{B}_{\mathcal{T}_1}\ \boldsymbol{A}^{K-3}\boldsymbol{B}_{\mathcal{T}_2}\ \ldots\ \boldsymbol{B}_{\mathcal{T}_{K-1}}\right]\right\} &(4.52)\\ &\geq \text{rank}\left\{\left[\boldsymbol{A}^{K-2}\boldsymbol{B}_{\mathcal{T}_1}\ \boldsymbol{A}^{K-3}\boldsymbol{B}_{\mathcal{T}_2}\ \ldots\ \boldsymbol{B}_{\mathcal{T}_{K-1}}\right]\right\} &(4.53)\\ &= \text{rank}\left\{\boldsymbol{\Phi}^{(K-1,s)}\right\}. &(4.54)\end{aligned}$$

However, if (4.49) is not true, we arrive at $\text{rank}\left\{\boldsymbol{\Phi}^{(K,s)}\right\} = \text{rank}\left\{\boldsymbol{\Phi}^{(K-1,s)}\right\}$. Also, the above relation holds for any $\mathcal{S}$, i.e.,

$$\text{ColSpace}\left\{\boldsymbol{A}^{K-1}\boldsymbol{B}_{\mathcal{S}}\right\} \subset \text{ColSpace}\left\{\left[\boldsymbol{A}^{K-2}\boldsymbol{B}_{\mathcal{T}_1}\ \boldsymbol{A}^{K-3}\boldsymbol{B}_{\mathcal{T}_2}\ \ldots\ \boldsymbol{B}_{\mathcal{T}_{K-1}}\right]\right\}. \tag{4.55}$$

Therefore, we derive

$$\text{ColSpace}\left\{\boldsymbol{A}^{K-1}\boldsymbol{B}\right\} \subset \text{ColSpace}\left\{\left[\boldsymbol{A}^{K-2}\boldsymbol{B}_{\mathcal{T}_1}\ \boldsymbol{A}^{K-3}\boldsymbol{B}_{\mathcal{T}_2}\ \ldots\ \boldsymbol{B}_{\mathcal{T}_{K-1}}\right]\right\}. \tag{4.56}$$

Consequently, we deduce that

$$\text{ColSpace}\left\{\boldsymbol{A}^{K_s^*-1}\boldsymbol{B}\right\} \subset \text{ColSpace}\left\{\left[\boldsymbol{A}^{K_s^*-2}\boldsymbol{B}_{\mathcal{T}_1}\ \boldsymbol{A}^{K_s^*-3}\boldsymbol{B}_{\mathcal{T}_2}\ \ldots\ \boldsymbol{A}^{K_s^*-K}\boldsymbol{B}_{\mathcal{T}_{K-1}}\right]\right\}. \tag{4.57}$$

In contrast, (4.50) implies that, for any $\mathcal{S}_k \subseteq \{1, 2, \ldots, m\}$ with $|\mathcal{S}_k| \leq s$, for $k = 1, 2, \ldots, K_s^* - 1$, we get

$$\text{ColSpace}\left\{\boldsymbol{A}^{K_s^*-1}\boldsymbol{B}\right\} \subsetneq \text{ColSpace}\left\{\left[\boldsymbol{A}^{K_s^*-2}\boldsymbol{B}_{\mathcal{S}_1}\ \boldsymbol{A}^{K_s^*-3}\boldsymbol{B}_{\mathcal{S}_2}\ \ldots\ \boldsymbol{B}_{\mathcal{S}_{K_s^*-1}}\right]\right\}. \tag{4.58}$$

Hence, our assumption that (4.49) does not hold leads to falsehood, and the proof is complete. □

Finally, we bound the sparse controllability index from above and below using Lemmas 4.1 and 4.2.

Theorem 4.7 *The s-sparse controllability index K_s^* defined in* (4.44) *of a controllable linear dynamical system* ($\boldsymbol{A} \in \mathbb{R}^{n\times n}, \boldsymbol{B} \in \mathbb{R}^{n\times m}$) *in* (4.1) *is bounded as*

$$\frac{n}{\min\{\text{rank}\left\{\boldsymbol{B}\right\}, s\}} \leq K^* \leq n - \min\{\text{rank}\left\{\boldsymbol{B}\right\}, s\} + 1. \tag{4.59}$$

Proof The lower bound directly follows from Lemma 4.1. Next, to prove the upper bound, using the definition in Lemma 4.2, we deduce that

$$n = \operatorname{rank}\left\{\boldsymbol{\Phi}^{(K_s^*,s)}\right\} \geq \operatorname{rank}\left\{\boldsymbol{\Phi}^{(K_s^*-1,s)}\right\} + 1 \geq \ldots \geq \operatorname{rank}\left\{\boldsymbol{\Phi}^{(1,s)}\right\} + K_s^* - 1 \tag{4.60}$$

$$= \max_{\substack{\mathcal{S}\subseteq\{1,2,\ldots,m\}\\ |\mathcal{S}|\leq s}} \operatorname{rank}\{\boldsymbol{B}_{\mathcal{S}}\} + K_s^* - 1 = \min\{\operatorname{rank}\{\boldsymbol{B}\}, s\} + K_s^* - 1. \tag{4.61}$$

Rearranging the above relation, we arrive at the upper bound. □

Theorem 4.7 implies that as s increases from 1 to $\operatorname{rank}\{\boldsymbol{B}\}$, the upper and lower bounds decrease. This trend indicates that the system has more flexibility and thus requires fewer input vectors to ensure controllability. When $s = \operatorname{rank}\{\boldsymbol{B}\}$, the above result reduces to the bounds for (unconstrained) controllability index given by Theorem 4.2. As presented next, the upper and lower bounds meet for some special cases.

Corollary 4.4 *The s-sparse controllability index K_s^* defined in* (4.44) *of a controllable linear dynamical system* ($\boldsymbol{A} \in \mathbb{R}^{n\times n}$, $\boldsymbol{B} \in \mathbb{R}^{n\times m}$) *in* (4.1) *satisfies $K_1^* = n$. Also, if* $\operatorname{rank}\{\boldsymbol{B}\} = n$, *we have* $K_n^* = 1$.

4.5 Summary

- A linear dynamical system is s-sparse controllable if it is controllable using s-sparse inputs.
- If a linear dynamical system is s-sparse controllable, there exists an actuator schedule independent of the initial and final states, which can guarantee the controllability of the system.
- A linear dynamical system is s-sparse controllable if and only if it is controllable and the sparsity level exceeds the nullity of the state matrix.
- An s-sparse controllable linear dynamical system ($\boldsymbol{A} \in \mathbb{R}^{n\times n}$, $\boldsymbol{B} \in \mathbb{R}^{n\times m}$) needs at least $\frac{n}{\min\{\operatorname{rank}\{\boldsymbol{B}\},s\}}$ and at most $n - \min\{\operatorname{rank}\{\boldsymbol{B}\}, s\} + 1$ control inputs to ensure controllability.

Notes

The results of classical controllability theory, their proofs, and detailed discussion can be found in Chen (1984). The sparse controllability-related results discussed in this chapter are from Joseph and Murthy (2020). The paper also discusses several other results which we omit here.

References

Chen CT (1984) Linear system theory and design. Saunders College publishing

Joseph G, Murthy CR (2020) Controllability of linear dynamical systems under input sparsity constraints. IEEE Trans Autom Control 66(2):924–931

Chapter 5
Design of Sparse Inputs with Time-Varying Support

Abstract This chapter introduces the problem of designing or estimating sparse inputs (with time-varying support) for a linear dynamical system. We discuss two settings: state-dependent and state-independent sparse input design. In the state-dependent case, we explore several sparsity-aware algorithms from the compressed sensing literature to tackle the sparse control problem. For the state-independent case, we present a sparse actuator scheduling approach using greedy algorithms.

Keywords State-dependent actuator scheduling · Piecewise sparsity · State-independent actuator scheduling · Sparse input estimation · Controllability metrics · Greedy scheduling

5.1 Sparse Input Design Problems

We recall the state space model of the linear dynamical system $(\boldsymbol{A}, \boldsymbol{B})$ with state $\boldsymbol{x}_k \in \mathbb{R}^n$ and input $\boldsymbol{u}_k \in \mathbb{R}^m$,

$$\boldsymbol{x}_k = \boldsymbol{A}\boldsymbol{x}_{k-1} + \boldsymbol{B}\boldsymbol{u}_k, \tag{5.1}$$

for discrete-time indices $k = 1, 2, \ldots$. Here, we impose the constraint that the inputs are sparse $\|\boldsymbol{u}_k\|_0 \leq s \ll m$, for $k = 1, 2 \ldots$. The goal here is to estimate the sparse inputs given the initial state $\boldsymbol{x}_0$ and the desired final state $\boldsymbol{x}_{\mathrm{f}}$. Specifically, the inputs $\{\boldsymbol{u}_k,\ k = 1, 2, \ldots, K\}$ should steer the system from the state $\boldsymbol{x}_0$ to $\boldsymbol{x}_K = \boldsymbol{x}_{\mathrm{f}}$, for some finite K. This problem is feasible if the linear dynamical system is controllable and the sparsity level s exceeds the nullity of $\boldsymbol{A}$, as given by Theorem 4.6. So, throughout this chapter, assume that the above two conditions hold.

From Theorem 4.7, we note that a system needs at most n sparse inputs to drive its state from any initial state to any final state. So, for a given initial state $\boldsymbol{x}_0 \in \mathbb{R}^n$ and final state $\boldsymbol{x}_{\mathrm{f}} \in \mathbb{R}^n$ the state-dependent actuator schedule-based design solves for n sparse inputs $\{\boldsymbol{u}_k\}_{k=1}^n$ such that

G. Joseph and C. R. Murthy, *Sparsity-Constrained Linear Dynamical Systems*, Springer Tracts in Electrical and Electronics Engineering, https://doi.org/10.1007/978-981-97-7090-8_5

$$\boldsymbol{x}_{\mathrm{f}} - \boldsymbol{A}^n \boldsymbol{x}_0 = \sum_{k=1}^{n} \boldsymbol{A}^{n-k} \boldsymbol{B} \boldsymbol{u}_k. \tag{5.2}$$

which follows from (5.1).

Let the vector of sparse unknowns be

$$\boldsymbol{u} = \begin{bmatrix} \boldsymbol{u}_1 \\ \boldsymbol{u}_2 \\ \vdots \\ \boldsymbol{u}_n \end{bmatrix}. \tag{5.3}$$

Then, we rewrite (5.2) as

$$\bar{\boldsymbol{x}} = \boldsymbol{x}_{\mathrm{f}} - \boldsymbol{A}^n \boldsymbol{x}_0 = \sum_{k=1}^{n} \boldsymbol{A}^{n-k} \boldsymbol{B} \boldsymbol{u}_k = \boldsymbol{\Phi} \boldsymbol{u}, \tag{5.4}$$

where we define the matrix $\boldsymbol{\Phi}$ as

$$\boldsymbol{\Phi} = \left[\boldsymbol{A}^{n-1} \boldsymbol{B} \ \boldsymbol{A}^{n-2} \boldsymbol{B} \ \ldots \ \boldsymbol{B} \right]. \tag{5.5}$$

Hence, solving for sparse inputs is the same as solving the system of equations

$$\bar{\boldsymbol{x}} = \boldsymbol{\Phi} \boldsymbol{u}, \tag{5.6}$$

for a sparse solution $\boldsymbol{u}$.

We divide the sparse input design algorithms for the time-varying support case into two categories: state-dependent actuator scheduling and state-independent actuator scheduling, as presented next.

5.2 State-Dependent Actuator Scheduling

The goal of the state-dependent actuator scheduling-based approach is to design n sparse inputs $\{\boldsymbol{u}_k\}_{k=1}^n$ by directly solving (5.6). We note that this approach does not assume that the system is sparse controllable, but it assumes (5.6) has a solution. In other words, this approach assumes that $\bar{\boldsymbol{x}}$ belongs to ColSpace $\{\boldsymbol{\Phi}\}$, which need not be equal to $\mathbb{R}^n$.

There are two approaches to finding a sparse solution to (5.6). In the first approach, we minimize the average sparsity of the inputs given by

$$\frac{1}{n} \sum_{k=1}^{n} \|\boldsymbol{u}_k\|_0 = \frac{1}{n} \|\boldsymbol{u}\|_0. \tag{5.7}$$

Therefore, by dropping the constant factor of $1/n$ from the objective function, the sparse input design problem is

$$\arg\min_{\boldsymbol{u}\in\mathbb{R}^{nm}} \quad \|\boldsymbol{u}\|_0 \quad \text{s. t.} \quad \bar{\boldsymbol{x}} = \boldsymbol{\Phi}\boldsymbol{u}. \tag{5.8}$$

The above problem is the same as the classical sparse signal recovery or compressed sensing problem. Hence, we can use the sparsity-aware algorithms from the compressed sensing literature to solve the problem. This approach is referred to as *sparse input design via sparse recovery algorithms*. Some of standard algorithms, such as convex optimization (Algorithm 2.1), greedy (Algorithms 2.2 and 2.3), thresholding(Algorithms 2.4 and 2.5), and Bayesian methods (Algorithm 2.6), to solve (5.8) are discussed in Chapter 2.

The problem with the above approach is that it is possible to have different sparsity levels for inputs at different times. For example, a solution can contain an input $\boldsymbol{u}_k$ with all nonzero entries because there is no control over the sparsity of the individual inputs. To tackle this problem, in the second approach, the sparsity level on each input is limited by a constant $s < m$ as follows:

$$\arg\min_{\boldsymbol{u}\in\mathbb{R}^{nm}} \ \|\boldsymbol{u}\|_0 \ \text{ s. t. } \bar{\boldsymbol{x}} = \boldsymbol{\Phi}\boldsymbol{u}, \ \boldsymbol{u} = \begin{bmatrix} \boldsymbol{u}_1 \\ \boldsymbol{u}_2 \\ \vdots \\ \boldsymbol{u}_n \end{bmatrix} \ \text{ and } \ \|\boldsymbol{u}_k\|_0 \leq s, \quad \text{for} \ \ k = 1, 2, \ldots, n. \tag{5.9}$$

Due to the additional constraint, the above optimization problem seeks $\boldsymbol{u}$ obtained by concatenating n vectors, each of which is s-sparse. The feasibility set of (5.9) is a subset of the feasibility set of (5.6). This special type of sparse vector with an additional structure is called the piecewise sparse vector, and this approach in (5.9) is referred to as *sparse input design via piecewise sparse recovery algorithms*.

5.3 Design via Piecewise Sparse Recovery Algorithms

We present two algorithms, variants of orthogonal matching pursuit and compressive sampling matching pursuit algorithms, that solve (5.9).

5.3.1 Piecewise Orthogonal Matching Pursuit

The piecewise orthogonal matching pursuit algorithm is similar to the classical one. The only difference between the two algorithms is that the piecewise orthogonal matching pursuit algorithm ensures that $\|\boldsymbol{u}_k\|_0 \leq s$ holds for all values of k by not adding elements to $\mathcal{U}$ corresponding to $\boldsymbol{u}_k$ if $\|\boldsymbol{u}_k\|_0 = s$. To this end, piecewise orthogonal matching pursuit tracks the time index k for which $\mathcal{S}_k = s$ and excludes them from the subsequent iterations. This set of "abandoned" values of k is given by the set $\{k \in \{1, 2, \ldots, n\} : |\mathcal{S}_k| = s\}$. When this set is empty, i.e., $|\mathcal{S}_k| < s$ for $k = 1, 2, \ldots, n$, piecewise orthogonal matching pursuit behaves like the regular orthogonal matching pursuit algorithm in Algorithm 2.2. The algorithm is summarized in Algorithm 5.1. We note that piecewise orthogonal matching pursuit is not the same as applying orthogonal matching pursuit n times because the vector $\bar{\boldsymbol{x}}$ is common to all the inputs $\boldsymbol{u}_k$'s.

Algorithm 5.1 Sparse input design via piecewise orthogonal matching pursuit

Input: Initial state $\boldsymbol{x}_0 \in \mathbb{R}^n$, final state $\boldsymbol{x}_\mathrm{f} \in \mathbb{R}^n$, linear dynamical system ($\boldsymbol{A} \in \mathbb{R}^{n\times n}$, $\boldsymbol{B} \in \mathbb{R}^{n\times m}$), sparsity level $s < m$

Initialization: $\mathcal{U} = \mathcal{S}_k = \emptyset$, for $k = 1, 2, \ldots, n$, $\boldsymbol{u} = \boldsymbol{0}$

1: Set $\bar{\boldsymbol{x}} = \boldsymbol{x}_\mathrm{f} - \boldsymbol{A}^n\boldsymbol{x}_0$ and $\boldsymbol{\Phi} = \left[\boldsymbol{A}^{n-1}\boldsymbol{B}\ \boldsymbol{A}^{n-2}\boldsymbol{B}\ \ldots\ \boldsymbol{B}\right]$

2: **repeat**

3: Compute $(k^*, i^*) = \underset{\substack{k:|\mathcal{S}_k|<s \\ i\notin\mathcal{S}_k}}{\arg\max} \dfrac{\left|\boldsymbol{\Phi}_{(k-1)m+i}^\mathsf{T}(\bar{\boldsymbol{x}}-\boldsymbol{\Phi}\boldsymbol{u})\right|}{\left\|\boldsymbol{\Phi}_{(k-1)m+i}\right\|}$

4: Update the support set of $\boldsymbol{u}_{k^*}$ as $\mathcal{S}_{k^*} \leftarrow \mathcal{S}_{k^*} \cup \{i^*\}$

5: Update the support set of $\boldsymbol{u}$ as $\mathcal{U} \leftarrow \mathcal{U} \cup \{(k^*-1)m + i^*\}$

6: Determine $\boldsymbol{u}_\mathcal{U} \leftarrow \boldsymbol{\Phi}_\mathcal{U}^\dagger\bar{\boldsymbol{x}}$ and $\boldsymbol{u}_{\mathcal{U}^\complement} \leftarrow \boldsymbol{0}$

7: **until** $\bar{\boldsymbol{x}} = \boldsymbol{\Phi}\boldsymbol{u}$

Output: Control inputs $\boldsymbol{u}$

5.3.2 *Piecewise Compressive Sampling Matching Pursuit*

Similar to piecewise orthogonal matching pursuit, we can modify the compressive sampling matching pursuit to get the piecewise compressive sampling matching pursuit algorithm. Instead of choosing the top $2ns$ indices in Step 3, the piecewise counterpart chooses the top $2s$ corresponding to each of the n time indices,

$$\mathcal{I}_k^* = \underset{\substack{\mathcal{I}\subseteq\{1,2,\ldots,m\} \\ |\mathcal{I}|\leq 2s}}{\arg\max} \sum_{i\in\mathcal{I}} \frac{\left|\left[\boldsymbol{\Phi}^\mathsf{T}(\bar{\boldsymbol{x}} - \boldsymbol{\Phi}\boldsymbol{u})\right]_{(k-1)m+i}\right|}{\left\|\boldsymbol{\Phi}_{(k-1)m+i}\right\|}. \tag{5.10}$$

Likewise, the top ns indices in Step 6 are replaced with the top s indices corresponding to each time index,

$$\mathcal{S}_k = \underset{\substack{\mathcal{I}\subseteq\{1,2,\ldots,m\} \\ |\mathcal{I}|\leq s}}{\arg\max} \sum_{i\in\mathcal{I}} \left|\boldsymbol{u}_{(k-1)m+i}\right|. \tag{5.11}$$

In other words, the support index identification steps are replaced by piecewise support identification to ensure the extra constraint: $\|\boldsymbol{u}_k\|_0 \leq s$ in (5.9). The pseudocode of the algorithm is summarized in Algorithm 5.2.

Algorithm 5.2 Sparse input design via piecewise compressive sampling matching pursuit

---

Input: Initial state $\boldsymbol{x}_0 \in \mathbb{R}^n$, final state $\boldsymbol{x}_\mathrm{f} \in \mathbb{R}^n$, linear dynamical system ($\boldsymbol{A} \in \mathbb{R}^{n\times n}$, $\boldsymbol{B} \in \mathbb{R}^{n\times m}$), sparsity level $s < m$

Initialization: $\mathcal{S}_k = \emptyset$, for $k = 1, 2, \ldots, n$, $\boldsymbol{u} = \boldsymbol{0}$

Set $\bar{\boldsymbol{x}} = \boldsymbol{x}_\mathrm{f} - \boldsymbol{A}^n\boldsymbol{x}_0$ and $\boldsymbol{\Phi} = \left[\boldsymbol{A}^{n-1}\boldsymbol{B}\ \boldsymbol{A}^{n-2}\boldsymbol{B}\ \ldots\ \boldsymbol{B}\right]$

repeat

  Update the support set for $k = 1, 2, \ldots, n$ as $\mathcal{S}_k \leftarrow \mathcal{S}_k \cup \mathcal{I}_k^*$ using

$$\mathcal{I}_k^* = \underset{\substack{\mathcal{I}\subseteq\{1,2,\ldots,m\}\\ |\mathcal{I}|\leq 2s}}{\arg\max} \sum_{i\in\mathcal{I}} \frac{\left|\boldsymbol{\Phi}_{(k-1)m+i}^\mathsf{T}(\bar{\boldsymbol{x}} - \boldsymbol{\Phi}\boldsymbol{u})\right|}{\left\|\boldsymbol{\Phi}_{(k-1)m+i}\right\|}$$

  Update the support set of $\boldsymbol{u}$ as $\mathcal{U} = \bigcup_{k=1}^n \{i + (k-1)m : i \in \mathcal{S}_k\}$

  Determine $\boldsymbol{u}_\mathcal{U} \leftarrow \boldsymbol{\Phi}_\mathcal{U}^\dagger \bar{\boldsymbol{x}}$ and $\boldsymbol{u}_{\mathcal{U}^\mathrm{C}} \leftarrow \boldsymbol{0}$

  Determine new support for $k = 1, 2, \ldots, n$ as

$$\mathcal{S}_k \leftarrow \underset{\substack{\mathcal{I}\subseteq\{1,2,\ldots,m\}\\ |\mathcal{I}|\leq s}}{\arg\max} \sum_{i\in\mathcal{I}} \left|\boldsymbol{u}_{(k-1)m+i}\right|$$

  Update the support set of $\boldsymbol{u}$ as $\mathcal{U} = \bigcup_{k=1}^n \{i + (k-1)m : i \in \mathcal{S}_k\}$

  Update the estimate $\boldsymbol{u}_{\mathcal{U}^\mathrm{C}} \leftarrow \boldsymbol{0}$

until $\bar{\boldsymbol{x}} = \boldsymbol{\Phi}\boldsymbol{u}$

Output: Control inputs $\boldsymbol{u}$

5.4 Sparse Input Estimation from Observations

It is worth noting that the sparse input design problem and algorithms are closely related to the sparse input estimation problem. The observation $\boldsymbol{y}_k \in \mathbb{R}^p$ at time $k = 1, 2, \ldots, n$ is

$$\boldsymbol{y}_k = \boldsymbol{C}\boldsymbol{x}_k, \tag{5.12}$$

where $\boldsymbol{C} \in \mathbb{R}^{p\times n}$ is the observation matrix. Assuming the initial state $\boldsymbol{x}_0$ to be known, the input estimation problem simplifies to

$$\bar{y} = \begin{bmatrix} y_1 \\ y_2 \\ \vdots \\ y_n \end{bmatrix} - \begin{bmatrix} CA \\ CA^2 \\ \vdots \\ CA^n \end{bmatrix} x_0 \tag{5.13}$$

$$= \begin{bmatrix} CB & & & & \\ CAB & CB & & & \\ CAB & CAB & CB & & \\ \vdots & \vdots & \vdots & \ddots & \\ CA^{n-1}B & CA^{n-2}B & CA^{n-3}B & \dots & CB \end{bmatrix} \begin{bmatrix} u_1 \\ u_2 \\ \vdots \\ u_n \end{bmatrix} \tag{5.14}$$

$$= \Psi u, \tag{5.15}$$

where u is defined in (5.3), and $\Psi \in \mathbb{R}^{np \times nm}$ is the corresponding measurement matrix. Hence, solving for sparse inputs is the same as finding the sparse solution u to the linear system of equations: $\bar{y} = \Psi u$. Therefore, all the algorithms (Algorithms 2.1–2.3, 5.1, and 5.2) can be used to estimate the sparse control inputs.

5.5 State-Independent Actuator Scheduling

The goal of sparse input design with state-independent actuator schedule is to find a sparse actuator schedule $\{\mathcal{S}_k \subseteq \{1, 2, \dots, m\}\}_{k=1}^n$ with $|\mathcal{S}_k| \leq s$ such that the resulting controllability matrix has full rank. The actuator schedule gives the indices of the nonzero entries $\mathcal{U}$ of u in (5.3), i.e.,

$$\mathcal{U} = \bigcup_{k=1}^{n} \{i + (k-1)m : i \in \mathcal{S}_k\}. \tag{5.16}$$

Therefore, the resulting controllability matrix is

$$\Phi_{\mathcal{U}} = \left[A^{n-1} B_{\mathcal{S}_1} \; A^{n-2} B_{\mathcal{S}_2} \; \dots \; B_{\mathcal{S}_n} \right], \tag{5.17}$$

and the desired condition is

$$\operatorname{rank}\{\Phi_{\mathcal{U}}\} = n. \tag{5.18}$$

Therefore, we derive $u_{\mathcal{U}^{\complement}} = \mathbf{0}$, and the system of equations (5.6) simplifies as

$$\bar{x} = \Phi_{\mathcal{U}} u_{\mathcal{U}}, \tag{5.19}$$

which does not have any sparsity constraint on the solution. Hence, for a given pair of initial and final states (x_0, x_f) is given by the nonzero entries of u are given by the least-square solution of (5.19),

$$\boldsymbol{u}_{\mathcal{U}} = \boldsymbol{\Phi}_{\mathcal{U}}^{\dagger}\bar{\boldsymbol{x}} \quad \text{and} \quad \boldsymbol{u}_{\mathcal{U}^{\complement}} = \boldsymbol{0}. \tag{5.20}$$

In the sequel, we discuss how to find a sparse actuator schedule and given the actuator schedule, the control inputs are given by (5.20).

As discussed in Sect. 5.2, we can solve for a sparse schedule using the following average sparsity constraint.

$$\underset{\substack{\mathcal{S}_k\subseteq\{1,2,\ldots,m\}\\ k=1,2,\ldots,n}}{\arg\min} \sum_{k=1}^{n} |\mathcal{S}_k| \text{ s. t. } \operatorname{rank}\left\{\left[\boldsymbol{A}^{n-1}\boldsymbol{B}_{\mathcal{S}_1}\ \boldsymbol{A}^{n-2}\boldsymbol{B}_{\mathcal{S}_2}\ \ldots\ \boldsymbol{B}_{\mathcal{S}_n}\right]\right\} = n. \tag{5.21}$$

Alternatively, we can solve for a sparse schedule by imposing a sparsity constraint on each control input $\boldsymbol{u}_k$, i.e.,

$$\underset{\substack{\mathcal{S}_k\subseteq\{1,2,\ldots,m\}\\ k=1,2,\ldots,n}}{\arg\min} \sum_{k=1}^{n} |\mathcal{S}_k| \quad \text{s. t. } |\mathcal{S}_k| \leq s, \quad \text{for } k = 1, 2, \ldots, n,$$
$$\operatorname{rank}\left\{\left[\boldsymbol{A}^{n-1}\boldsymbol{B}_{\mathcal{S}_1}\ \boldsymbol{A}^{n-2}\boldsymbol{B}_{\mathcal{S}_2}\ \ldots\ \boldsymbol{B}_{\mathcal{S}_n}\right]\right\} = n. \tag{5.22}$$

The above combinatorial problems are hard to solve, and unlike the state-dependent sparse actuator scheduling problem, we do not optimize the control input here. Also, the solution to the problem may not be unique, and the different solution schedules can require different energy for the control inputs to drive the system to a desired state. For example, from (5.20), for a given controllability Gramian matrix $\boldsymbol{W} = \boldsymbol{\Phi}_{\mathcal{U}}\boldsymbol{\Phi}_{\mathcal{U}}^{\mathsf{T}}$, the worst-case energy of the control inputs is given by

$$\max_{\bar{\boldsymbol{x}}:\|\bar{\boldsymbol{x}}\|=1} \left\|\boldsymbol{\Phi}_{\mathcal{U}}^{\dagger}\bar{\boldsymbol{x}}\right\|^2 = \max_{\bar{\boldsymbol{x}}:\|\bar{\boldsymbol{x}}\|=1} \bar{\boldsymbol{x}}^{\mathsf{T}}\boldsymbol{W}^{-1}\bar{\boldsymbol{x}} = \lambda_{\max}\left(\boldsymbol{W}^{-1}\right) = \frac{1}{\lambda_{\min}\left(\boldsymbol{W}\right)}, \tag{5.23}$$

where $\lambda_{\max}$ and $\lambda_{\min}$ denote a matrix's maximum and minimum absolute eigenvalues, respectively. So, one approach to solving for a sparse schedule is to minimize the inverse of the minimum eigenvalue of the resulting controllability matrix. Similarly, we can design other controllability metrics to quantify the required control input energy in different ways. Table 5.1 summarizes a few popular control metrics from the literature.

Table 5.1 Some popular controllability metrics, with $\boldsymbol{W}$ denoting the resulting controllability Gramian

Controllability measure	Controllability metric
Average control energy	$\operatorname{tr}\left\{\boldsymbol{W}^{-1}\right\}$
The volume of the ellipsoid	$\|\boldsymbol{W}\|^{-1/n}$
Inverse of the trace	$\operatorname{tr}\{\boldsymbol{W}\}^{-1}$
Worst-case control energy	$1/\lambda_{\min}(\boldsymbol{W})$

Based on the above observations, we reformulate the sparse scheduling problem in (5.21) as follows:

$$\underset{\substack{\mathcal{S}_k \subseteq \{1,2,\ldots,m\} \\ k=1,2,\ldots,n}}{\arg\min} \; \rho(\boldsymbol{W}) \text{ s. t. } \sum_{k=1}^{n} |\mathcal{S}_k| \leq ns, \quad \text{rank}\{\boldsymbol{W}\} = n, \quad \boldsymbol{W} = \boldsymbol{\Phi}_{\mathcal{U}} \boldsymbol{\Phi}_{\mathcal{U}}^{\mathsf{T}},$$
$$\text{and } \boldsymbol{\Phi}_{\mathcal{U}} = \left[\boldsymbol{A}^{n-1}\boldsymbol{B}_{\mathcal{S}_1} \; \boldsymbol{A}^{n-2}\boldsymbol{B}_{\mathcal{S}_2} \; \ldots \; \boldsymbol{B}_{\mathcal{S}_n}\right], \tag{5.24}$$

where ρ can be any controllability metric (see Table 5.1 for examples). Moreover, we can rewrite $\boldsymbol{W}$ in the above equation as

$$\boldsymbol{W} = \sum_{k=1}^{n} \sum_{i \in \mathcal{S}_k} \boldsymbol{A}^{n-k} \boldsymbol{B}_i \boldsymbol{B}_i^{\mathsf{T}} \boldsymbol{A}^{(n-k)\mathsf{T}}. \tag{5.25}$$

As a result, (5.24) is equivalent to the following optimization problem:

$$\underset{\substack{\mathcal{S}_k \subseteq \{1,2,\ldots,m\} \\ k=1,2,\ldots,n}}{\arg\min} \; \rho\left(\sum_{k=1}^{n} \sum_{i \in \mathcal{S}_k} \boldsymbol{A}^{n-k} \boldsymbol{B}_i \boldsymbol{B}_i^{\mathsf{T}} \boldsymbol{A}^{(n-k)\mathsf{T}}\right) \text{ s. t. } \sum_{k=1}^{n} |\mathcal{S}_k| \leq ns. \tag{5.26}$$

Similarly, we can formulate the sparse scheduling problem with the sparsity constraint on each control input as follows:

$$\underset{\substack{\mathcal{S}_k \subseteq \{1,2,\ldots,m\} \\ k=1,2,\ldots,n}}{\arg\min} \; \rho\left(\sum_{k=1}^{n} \sum_{i \in \mathcal{S}_k} \boldsymbol{A}^{n-k} \boldsymbol{B}_i \boldsymbol{B}_i^{\mathsf{T}} \boldsymbol{A}^{(n-k)\mathsf{T}}\right) \text{ s. t. } |\mathcal{S}_k| \leq s, k = 1, 2, \ldots, n. \tag{5.27}$$

5.6 Sparse Actuator Scheduling Algorithms

The greedy heuristic sequentially selects one control input index in every iteration to maximize the controllability metric. Here, the controllability metrics like average control energy (tr $\left\{\boldsymbol{W}^{-1}\right\}$) and worst-case control energy ($1/\lambda_{\min}(\boldsymbol{W})$) are undefined if $\boldsymbol{W}$ is not invertible. So, to avoid singularity, we use a small constant $\epsilon > 0$ to ensure that $\boldsymbol{W} + \epsilon \boldsymbol{I}$ is invertible. So the greedy cost function is $\rho(\boldsymbol{W} + \epsilon \boldsymbol{I})$, where ρ is the controllability metric.

The algorithms are similar in spirit to the orthogonal matching pursuit and piecewise orthogonal matching pursuit. They add one control input index per iteration. Also, once an index is added, it is not removed in the subsequent iterations. The iterations are continued until the resulting controllability matrix is full rank. The resulting algorithm for the average sparsity problem in (5.26) is summarized in Algorithm 5.3. A similar greedy algorithm for the per-time instant sparsity-constrained problem in

(5.27) is summarized in Algorithm 5.4. This algorithm obeys the piecewise sparsity constraints while adding an index to the support.

Algorithm 5.3 Greedy sparse actuator scheduling

Input: Linear dynamical system ($\boldsymbol{A} \in \mathbb{R}^{n \times n}$, $\boldsymbol{B} \in \mathbb{R}^{n \times m}$)
Initialization: $\mathcal{U} = \emptyset$, $\boldsymbol{W} = \boldsymbol{0}$, $\epsilon > 0$
1: Compute $\boldsymbol{\Phi} = \left[\boldsymbol{A}^{n-1}\boldsymbol{B}\ \boldsymbol{A}^{n-2}\boldsymbol{B}\ \ldots\ \boldsymbol{B}\right]$
2: **repeat**
3: Compute $i^* = \arg\min_{i \notin \mathcal{U}} \rho\left(\boldsymbol{W} + \boldsymbol{\Phi}_i \boldsymbol{\Phi}_i^{\mathsf{T}} + \epsilon \boldsymbol{I}\right)$
4: Update support $\mathcal{U} \leftarrow \mathcal{U} \cup \{i^*\}$
5: Update the Gramian matrix $\boldsymbol{W} \leftarrow \boldsymbol{W} + \boldsymbol{\Phi}_{i^*} \boldsymbol{\Phi}_{i^*}^{\mathsf{T}}$
6: **until** rank $\{\boldsymbol{W}\} = n$
Output: Actuator schedule $\mathcal{U}$

Algorithm 5.4 Greedy piecewise sparse actuator scheduling algorithm

Input: Linear dynamical system ($\boldsymbol{A} \in \mathbb{R}^{n \times n}$, $\boldsymbol{B} \in \mathbb{R}^{n \times m}$), sparsity level $s < m$
Initialization: $\mathcal{S}_k = \emptyset$, for $k = 1, 2, \ldots, n$, $\boldsymbol{W} = \boldsymbol{0}$, $\epsilon > 0$
1: Set $\boldsymbol{\Phi} = \left[\boldsymbol{A}^{n-1}\boldsymbol{B}\ \boldsymbol{A}^{n-2}\boldsymbol{B}\ \ldots\ \boldsymbol{B}\right]$
2: **repeat**
3: Compute $(k^*, i^*) = \underset{k:|\mathcal{S}_k|<s,\, i \notin \mathcal{S}_k}{\arg\min}\ \rho\left(\boldsymbol{W} + \boldsymbol{\Phi}_{(k-1)m+i} \boldsymbol{\Phi}_{(k-1)m+i}^{\mathsf{T}} + \epsilon \boldsymbol{I}\right)$
4: Update the support set as $\mathcal{S}_{k^*} \leftarrow \mathcal{S}_{k^*} \cup \{i^*\}$
5: Update the Gramian matrix $\boldsymbol{W} \leftarrow \boldsymbol{W} + \boldsymbol{\Phi}_{(k^*-1)m+i^*} \boldsymbol{\Phi}_{(k^*-1)m+i^*}^{\mathsf{T}}$
6: **until** rank $\{\boldsymbol{W}\} = n$
Output: Actuator schedule $\{\mathcal{S}_k\}_{k=1}^{n}$

5.7 Summary

- For sparse controllable linear dynamical systems, sparse inputs can be designed using actuator scheduling that can either be dependent or independent of initial and final states.
- The state-dependent sparse actuator scheduling with average sparsity constraint uses the standard sparse signal recovery algorithms.
- The state-dependent sparse actuator scheduling with per-time instant sparsity constraint uses piecewise sparse signal recovery algorithms.
- The sparse input estimation from the observations of a linear dynamical system is closely related to the state-dependent sparse actuator scheduling problem and uses the same set of algorithms.
- The state-independent sparse actuator scheduling relies on suitable controllability metrics to also optimize the control energy and uses greedy algorithms to optimize the schedule.

Notes

Unlike the standard sparse recovery problem, the piecewise sparse recovery problem is not well-studied, and only limited greedy algorithms are available in the literature. The piecewise orthogonal matching pursuit is from Li et al. (2016), and the piecewise compressive sampling matching pursuit is from Zhong and Li (2018). The other similar algorithms and guarantees for piecewise sparse recovery can be found in Li et al. (2016); Zhong and Li (2018) and the papers that cite them. Other than the greedy heuristics, more efficient algorithms to solve (5.9) are still an open problem in the sparse signal processing literature. The algorithms and guarantees for sparse input estimation from observations and the other variants of the problem are discussed in Sefati et al. (2015); Kafashan et al. (2016); Fosson et al. (2019) and their references. Table 5.1 and the greedy sparse actuator scheduling algorithms are mostly from Siami et al. (2020). We note that Siami et al. (2020) has a more detailed discussion on the controllability metrics and more sophisticated formulation and algorithms to solve the sparse actuator scheduling problem. Extending the compressive sampling matching pursuit algorithms for state-independent actuator scheduling is non-trivial and is an open problem.

References

Fosson SM, Garin F, Gracy S, Kibangou AY, Swart D (2019) Input and state estimation exploiting input sparsity. In: Proceedings of european control conference, pp 2344–2349

Kafashan M, Nandi A, Ching S (2016) Relating observability and compressed sensing of time-varying signals in recurrent linear networks. Neural Netw 83:11–20

Li K, Rojas CR, Yang T, Hjalmarsson H, Johansson KH, Cong S (2016) Piecewise sparse signal recovery via piecewise orthogonal matching pursuit. In: Proceedings of IEEE international conference on acoustics, speech and signal processing, pp 4608–4612

Sefati S, Cowan NJ, Vidal R (2015) Linear systems with sparse inputs: observability and input recovery. In: Proceedings of American control conference, pp 5251–5257

Siami M, Olshevsky A, Jadbabaie A (2020) Deterministic and randomized actuator scheduling with guaranteed performance bounds. IEEE Trans Automc Control 66(4):1686–1701

Zhong J, Li C (2018) Piecewise sparse recovery via piecewise greedy method. J Math Res Appl 6:643–658

Chapter 6
Stabilizability, Output Controllability, and Nonnegative Controllablity

Abstract This chapter builds upon the concept of sparse controllability and presents three related notions of controllability under input sparsity constraints: stabilizability, output controllability, and nonnegative controllability. We present three main results. We first prove that sparse stabilizability is equivalent to standard stabilizability. Next, we show that if a linear dynamical system is output controllable, there exist upper and lower bounds on the minimum sparsity level required to ensure sparse output controllability. Finally, we establish that a linear dynamical system is sparse nonnegative controllable if and only if it is both nonnegative controllable and sparse controllable.

Keywords Sparse control inputs · Time-varying support · Jordan canonical decomposition · Non-negative sparse recovery

6.1 Sparse Stabilizability

We recall the state space model of the linear dynamical system $(\boldsymbol{A}, \boldsymbol{B})$, with state $\boldsymbol{x}_k \in \mathbb{R}^n$ and input $\boldsymbol{u}_k \in \mathbb{R}^m$,

$$\boldsymbol{x}_k = \boldsymbol{A}\boldsymbol{x}_{k-1} + \boldsymbol{B}\boldsymbol{u}_k, \tag{6.1}$$

for discrete-time indices $k = 1, 2, \ldots$. The linear dynamical system $(\boldsymbol{A}, \boldsymbol{B})$ in (6.1) is said to be (non-sparse) stabilizable if, for any initial state $\boldsymbol{x}_0$, there exists a sequence of inputs $\{\boldsymbol{u}_k\}_{k>0}$ such that $\lim_{k\to\infty} \|\boldsymbol{x}_k\| = 0$. The classical test for stabilizability is as follows:

Theorem 6.1 *The linear dynamical system* $(\boldsymbol{A} \in \mathbb{R}^{n\times n}, \boldsymbol{B} \in \mathbb{R}^{n\times m})$ *in* (6.1) *is stabilizable if and only if*

$$\text{rank}\left\{\left[\boldsymbol{A} - \lambda \boldsymbol{I} \;\; \boldsymbol{B}\right]\right\} = n, \tag{6.2}$$

for all $\lambda \in \mathbb{C}$ *such that* $|\lambda| \geq 1$.

G. Joseph and C. R. Murthy, *Sparsity-Constrained Linear Dynamical Systems*,
Springer Tracts in Electrical and Electronics Engineering,
https://doi.org/10.1007/978-981-97-7090-8_6

We extend the notion of stabilizability to the case of sparse inputs and derive the stabilizability test as follows.

Definition 6.1 The linear dynamical system $(\boldsymbol{A}, \boldsymbol{B})$ in (6.1) is said to be s-sparse stabilizable for a given $s \geq 1$ if, for any initial state $\boldsymbol{x}_0$, there exists a sequence of s-sparse inputs $\{\boldsymbol{u}_k : \|\boldsymbol{u}_k\|_0 \leq s\}_{k>0}$ such that $\lim_{k\to\infty} \|\boldsymbol{x}_k\| = 0$.

As in the case of sparse controllability, sparse stabilizability is a stronger notion than stabilizability and a weaker notion than s-sparse controllability. So, the conditions for stabilizability in Theorem 6.1 are necessary for sparse stabilizability, and the conditions for sparse controllability in Theorem 4.6 are sufficient for s-sparse stabilizability. Further, we recall that Theorem 6.1 asserts that the system is stabilizable if and only if the unstable part of the system is controllable. Therefore, the system is s-sparse stabilizable if and only if the unstable part of the system is s-sparse controllable. However, the eigenvalues of the state matrix corresponding to the unstable part of the system are greater than or equal to 1; hence, the unstable part is always s-sparse controllable. Therefore, we conclude that stabilizable systems are sparse stabilizable. This result is summarized below.

Theorem 6.2 *For any sparsity level, $s \geq 1$, the linear dynamical system $(\boldsymbol{A}, \boldsymbol{B})$ in (6.1) is s-sparse stabilizable if and only if it is stabilizable.*

Proof Since any sparse stabilizable linear dynamical system is also stabilizable, it is enough to show that stabilizability implies sparse stabilizability. To prove this claim, suppose that the system is stabilizable. We then construct the canonical form using the real Jordan decomposition of the state transition matrix $\boldsymbol{A}$ as follows:

$$\boldsymbol{A} = \boldsymbol{P}^{-1} \begin{bmatrix} \boldsymbol{S} & \boldsymbol{0} \\ \boldsymbol{0} & \boldsymbol{T} \end{bmatrix} \boldsymbol{P} = \begin{bmatrix} \boldsymbol{P}^{(1)} \\ \boldsymbol{P}^{(2)} \end{bmatrix}^{-1} \begin{bmatrix} \boldsymbol{S} & \boldsymbol{0} \\ \boldsymbol{0} & \boldsymbol{T} \end{bmatrix} \begin{bmatrix} \boldsymbol{P}^{(1)} \\ \boldsymbol{P}^{(2)} \end{bmatrix}, \tag{6.3}$$

where $\boldsymbol{P} = \begin{bmatrix} \boldsymbol{P}^{(1)\mathsf{T}} & \boldsymbol{P}^{(2)\mathsf{T}} \end{bmatrix}^{\mathsf{T}}$ is an invertible matrix, and the square matrices $\boldsymbol{S}$ and $\boldsymbol{T}$ are formed by the Jordan blocks of $\boldsymbol{A}$ corresponding to the eigenvalues whose absolute value is greater than 1 and the eigenvalues whose absolute value is less than 1, respectively.

Premultiplying (6.1) with $\boldsymbol{P}$ gives

$$\boldsymbol{P}^{(1)}\boldsymbol{x}_k = \boldsymbol{S}\boldsymbol{P}^{(1)}\boldsymbol{x}_{k-1} + \boldsymbol{P}^{(1)}\boldsymbol{B}\boldsymbol{u}_k \tag{6.4}$$

$$\boldsymbol{P}^{(2)}\boldsymbol{x}_k = \boldsymbol{T}\boldsymbol{P}^{(2)}\boldsymbol{x}_{k-1} + \boldsymbol{P}^{(2)}\boldsymbol{B}\boldsymbol{u}_k. \tag{6.5}$$

Let $\boldsymbol{x}_k^{(1)} = \boldsymbol{P}^{(1)}\boldsymbol{x}_k$ and $\boldsymbol{x}_k^{(2)} = \boldsymbol{P}^{(2)}\boldsymbol{x}_k$ for all values of $k \geq 0$. Then, we arrive at

$$\boldsymbol{x}_k^{(1)} = \boldsymbol{S}\boldsymbol{x}_{k-1}^{(1)} + \boldsymbol{P}^{(1)}\boldsymbol{B}\boldsymbol{u}_k \tag{6.6}$$

$$\boldsymbol{x}_k^{(2)} = \boldsymbol{T}\boldsymbol{x}_{k-1}^{(2)} + \boldsymbol{P}^{(2)}\boldsymbol{B}\boldsymbol{u}_k. \tag{6.7}$$

Then, we have two linear dynamical systems $(\boldsymbol{S}, \boldsymbol{P}^{(1)}\boldsymbol{B})$ and $(\boldsymbol{T}, \boldsymbol{P}^{(2)}\boldsymbol{B})$ with a common input $\boldsymbol{u}_k$.

Further, since the linear dynamical system $(\boldsymbol{A}, \boldsymbol{B})$ is stabilizable, Theorem 6.1 ensures that, for any $\lambda \in \mathbb{C}$ such that $|\lambda| \geq 1$,

$$\operatorname{rank}\left\{\left[\boldsymbol{A}-\lambda \boldsymbol{I} \; \boldsymbol{B}\right]\right\}=n. \tag{6.8}$$

Also, from (6.3), we obtain

$$n=\operatorname{rank}\left\{\boldsymbol{P}\left[\boldsymbol{A}-\lambda \boldsymbol{I} \; \boldsymbol{B}\right]\begin{bmatrix}\boldsymbol{P}^{-1} & \mathbf{0}\\ \mathbf{0} & \boldsymbol{I}\end{bmatrix}\right\} \tag{6.9}$$

$$=\operatorname{rank}\left\{\begin{bmatrix}\boldsymbol{S}-\lambda \boldsymbol{I} & \mathbf{0} & \boldsymbol{P}^{(1)}\boldsymbol{B}\\ \mathbf{0} & \boldsymbol{T}-\lambda \boldsymbol{I} & \boldsymbol{P}^{(2)}\boldsymbol{B}\end{bmatrix}\right\}. \tag{6.10}$$

Hence, $\left[\boldsymbol{S}-\lambda \boldsymbol{I} \; \boldsymbol{P}^{(1)}\boldsymbol{B}\right]$ has full row rank for any $\lambda \in \mathbb{C}$ such that $|\lambda| \geq 1$. Furthermore, from (6.3), the absolute value of all eigenvalues of $\boldsymbol{S}$ are greater than 1. So, $\boldsymbol{S}$ is invertible and $\left[\boldsymbol{S}-\lambda \boldsymbol{I} \; \boldsymbol{P}^{(1)}\boldsymbol{B}\right]$ has full row rank for any $\lambda \in \mathbb{C}$ such that $|\lambda| < 1$. In other words, $\left[\boldsymbol{S}-\lambda \boldsymbol{I} \; \boldsymbol{P}^{(1)}\boldsymbol{B}\right]$ has full row rank for any $\lambda \in \mathbb{C}$. From Corollary 4.3, we conclude that the linear dynamical system $(\boldsymbol{S}, \boldsymbol{P}^{(1)}\boldsymbol{B})$ is s-sparse controllable for any $s > 0$. Hence, for any initial state $\boldsymbol{x}_0$, we can find inputs s-sparse inputs $\{\boldsymbol{u}_k\}_{k=1}^{n}$ such that the linear dynamical system $(\boldsymbol{S}, \boldsymbol{P}^{(1)}\boldsymbol{B})$ is driven to the all-zero state, i.e., $\boldsymbol{x}_n^{(1)} = \boldsymbol{P}^{(1)}\boldsymbol{x}_n = \mathbf{0}$. However, these inputs do ensure that $\boldsymbol{P}^{(2)}\boldsymbol{x}_n = \mathbf{0}$.

Furthermore, if we choose $\boldsymbol{u}_k = 0$ for $k > n$, from (6.6), we get

$$\boldsymbol{P}^{(1)}\boldsymbol{x}_k=\mathbf{0} \quad \forall\; k>n. \tag{6.11}$$

Hence, from (6.7), for any $k > n$,

$$\|\boldsymbol{x}_k\| = \left\|\boldsymbol{P}^{-1}\boldsymbol{P}\boldsymbol{x}_k\right\| \leq \left\|\boldsymbol{P}^{-1}\right\| \left\|\boldsymbol{P}\boldsymbol{x}_k\right\| \tag{6.12}$$

$$= \left\|\boldsymbol{P}^{-1}\right\|\left(\left\|\boldsymbol{P}^{(1)}\boldsymbol{x}_k\right\| + \left\|\boldsymbol{P}^{(2)}\boldsymbol{x}_k\right\|\right) = \left\|\boldsymbol{P}^{-1}\right\| \left\|\boldsymbol{P}^{(2)}\boldsymbol{x}_k\right\|, \tag{6.13}$$

where we use (6.11). Using the fact that $\boldsymbol{u}_k = 0$ for $k > n$, we have

$$\boldsymbol{P}^{(2)}\boldsymbol{x}_k = \boldsymbol{T}\boldsymbol{P}^{(2)}\boldsymbol{x}_{k-1} = \boldsymbol{T}^{k-n}\boldsymbol{P}^{(2)}\boldsymbol{x}_n. \tag{6.14}$$

Consequently, we can further upper bound (6.13) as

$$\|\boldsymbol{x}_k\| \leq \left\|\boldsymbol{P}^{-1}\right\| \left\|\boldsymbol{T}^{k-n}\boldsymbol{P}^{(2)}\boldsymbol{x}_n\right\| \tag{6.15}$$

$$\leq \|\boldsymbol{T}\|^{k-n} \left\|\boldsymbol{P}^{-1}\right\| \left\|\boldsymbol{P}^{(2)}\boldsymbol{x}_n\right\|. \tag{6.16}$$

Since all the eigenvalues of $\boldsymbol{T}$ are less than 1, we deduce

$$\lim_{k\to\infty} \|\boldsymbol{x}_k\| = 0. \tag{6.17}$$

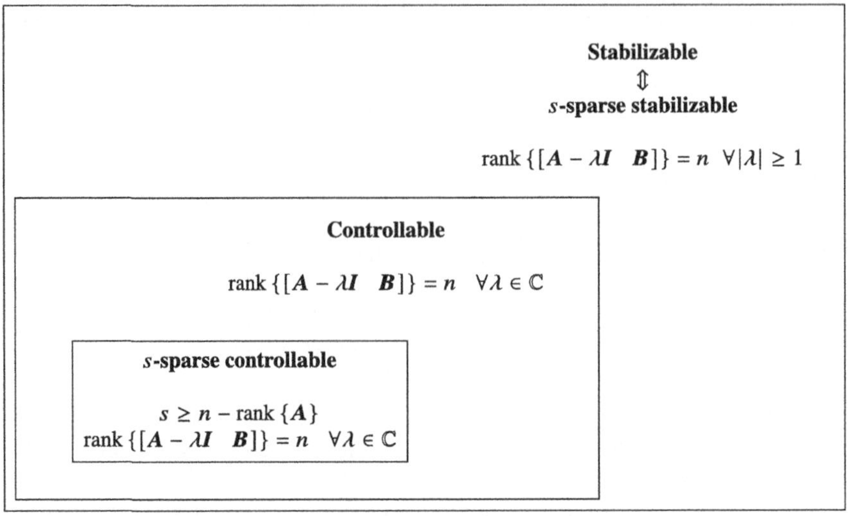

Fig. 6.1 Venn diagram showing the relation between sparse controllability, controllability, sparse stabilizability, and stabilizability, and the necessary and sufficient conditions each set satisfies

Hence, the linear dynamical system $(\boldsymbol{A}, \boldsymbol{B})$ is s-sparse stabilizable for any $s > 0$. The proof is complete. □

The above result establishes that sparse stabilizability is equivalent to stabilizability, unlike its controllability counterpart. The resulting relationship between sparse controllability, controllability, and sparse stabilizability is depicted by the Venn diagram in Fig. 6.1.

The proof of Theorem 6.2 provides an approach to finding sparse inputs that stabilize the system. Specifically, for any initial state $\boldsymbol{x}_0$, we determine sparse inputs $\{\boldsymbol{u}_k\}_{k=1}^n$ that drive the system $(\boldsymbol{S}, \boldsymbol{P}^{(1)}\boldsymbol{B})$ from $\boldsymbol{P}^{(1)}\boldsymbol{x}_0$ to the all-zero state. Here, $\boldsymbol{S}$ and $\boldsymbol{P}^{(1)}$ are defined in (6.3). The sparse inputs can be determined using any of the algorithms described in Chapters 2 and 5.

6.2 Sparse Output Controllability

To define the notation of output controllability, we consider a linear dynamical system whose state and observations are as follows:

$$\boldsymbol{x}_k = \boldsymbol{A}\boldsymbol{x}_{k-1} + \boldsymbol{B}\boldsymbol{u}_k \quad \text{and} \quad \boldsymbol{y}_k = \boldsymbol{C}\boldsymbol{x}_k, \tag{6.18}$$

where $\boldsymbol{y}_k$ is the observation at time k and $\boldsymbol{C}$ is the observation matrix. The linear dynamical system $(\boldsymbol{A}, \boldsymbol{B}, \boldsymbol{C})$ in (6.18) is said to be (non-sparse) output controllable if, for any initial state $\boldsymbol{x}_0$ and final output $\boldsymbol{y}_{\mathrm{f}}$, there exist inputs $\{\boldsymbol{u}_k\}_{k=1}^n$ that drive the system from $\boldsymbol{x}_0$ to the output $\boldsymbol{y}_n = \boldsymbol{y}_f$. The classical Kalman rank test for output controllability is as follows:

Theorem 6.3 *The linear dynamical system* ($\boldsymbol{A} \in \mathbb{R}^{n\times n}$, $\boldsymbol{B} \in \mathbb{R}^{n\times m}$, $\mathbb{C} \in \mathbb{R}^{p\times n}$) *in* (6.18) *is output controllable if and only if*

$$\text{rank}\left\{\boldsymbol{C}\left[\boldsymbol{A}^{n-1}\boldsymbol{B}\ \boldsymbol{A}^{n-2}\boldsymbol{B}\ \ldots\ \boldsymbol{B}\right]\right\} = p. \tag{6.19}$$

We extend the notion of output controllability to the case of sparse inputs and derive the necessary and sufficient conditions for output controllability as follows.

Definition 6.2 The linear dynamical system $(\boldsymbol{A}, \boldsymbol{B}, \boldsymbol{C})$ in (6.18) is said to be s-sparse output controllable for a given $s \geq 1$ if, for any initial state $\boldsymbol{x}_0$ and final output $\boldsymbol{y}_f$, there exist an integer $K < \infty$ and s-sparse inputs $\{\boldsymbol{u}_k : \|\boldsymbol{u}_k\|_0 \leq s\}_{k=1}^{K}$ that drive the system from $\boldsymbol{x}_0$ to the output $\boldsymbol{y}_n = \boldsymbol{y}_f$.

The extension of the Kalman rank test for sparse output controllability is as follows.

Theorem 6.4 *The linear dynamical system* ($\boldsymbol{A} \in \mathbb{R}^{n\times n}$, $\boldsymbol{B} \in \mathbb{R}^{n\times m}$, $\boldsymbol{C} \in \mathbb{R}^{p\times n}$) *in* (6.18) *is s-sparse output controllable if and only if there exist an integer $K < \infty$ and sets $\{\mathcal{S}_k \subseteq \{1, 2, \ldots, m\},\ |\mathcal{S}_k| \leq s\}_{k=1}^{K}$ such that*

$$\text{rank}\left\{\boldsymbol{C}\left[\boldsymbol{A}^{K-1}\boldsymbol{B}_{\mathcal{S}_1}\ \boldsymbol{A}^{K-2}\boldsymbol{B}_{\mathcal{S}_2}\ \ldots\ \boldsymbol{B}_{\mathcal{S}_K}\right]\right\} = p. \tag{6.20}$$

Proof The proof is similar to that of Theorem 4.4, and we omit it here. □

Although Theorem 6.4 gives the necessary and sufficient condition for sparse output controllability, verifying the condition is computationally expensive due to its combinatorial nature. Therefore, we seek a result similar to Theorem 4.6 for output controllability. However, Theorem 4.6 is based on the PBH test for controllability, and an analogous PBH test is not available for output sparse controllability. So, the extension of Theorem 4.6 to output sparse controllability is not straightforward and needs new proof techniques.

As in the case of sparse controllability, sparse output controllability is a stronger notion than output controllability. So, the condition $\text{rank}\{\boldsymbol{C}\boldsymbol{\Phi}\} = n$ for output controllability in Theorem 6.3 is necessary for sparse output controllability, where

$$\boldsymbol{\Phi} = \left[\boldsymbol{A}^{n-1}\boldsymbol{B}\ \boldsymbol{A}^{n-2}\boldsymbol{B}\ \ldots\ \boldsymbol{B}\right] \in \mathbb{R}^{n\times nm}. \tag{6.21}$$

We first provide weaker but non-combinatorial conditions to check sparse output controllability. The first result establishes the necessary conditions for sparse output controllability of a controllable system.

Proposition 6.1 *The linear dynamical system* ($\boldsymbol{A} \in \mathbb{R}^{n\times n}$, $\boldsymbol{B} \in \mathbb{R}^{n\times m}$, $\boldsymbol{C} \in \mathbb{R}^{p\times n}$) *in* (6.18) *is s-sparse output controllable only if*

$$\text{rank}\{\boldsymbol{C}\boldsymbol{\Phi}\} = p \text{ and } s \geq \max_{1\leq i\leq n} \frac{1}{i}\left(p - \text{rank}\left\{\boldsymbol{C}\boldsymbol{A}^{i}\right\}\right), \tag{6.22}$$

where $\boldsymbol{\Phi}$ *is the controllability matrix defined in* (6.21).

Proof The necessity of rank $\{\boldsymbol{C}\boldsymbol{\Phi}\} = p$ is trivial from Theorem 6.3. Next, if the linear dynamical system $(\boldsymbol{A}, \boldsymbol{B}, \boldsymbol{C})$ in (6.18) is s-sparse output controllable, from Theorem 6.4, there exist an integer $K < \infty$ and sets $\{\mathcal{S}_k \subseteq \{1, 2, \ldots, m\},\ |\mathcal{S}_k| = s\}_{k=1}^{K}$ such that

$$\operatorname{rank}\left\{\boldsymbol{C}\left[\boldsymbol{A}^{K-1}\boldsymbol{B}_{\mathcal{S}_1}\ \boldsymbol{A}^{K-2}\boldsymbol{B}_{\mathcal{S}_2}\ \ldots\ \boldsymbol{B}_{\mathcal{S}_K}\right]\right\} = \mathbb{R}^p. \tag{6.23}$$

Also, for any $1 \le i \le K-1$, we have

$$\operatorname{ColSpace}\left\{\boldsymbol{C}\left[\boldsymbol{A}^{K-1}\boldsymbol{B}_{\mathcal{S}_1}\ \boldsymbol{A}^{K-2}\boldsymbol{B}_{\mathcal{S}_2}\ \ldots\ \boldsymbol{A}^{i}\boldsymbol{B}_{\mathcal{S}_{K-i}}\right]\right\} \subseteq \operatorname{ColSpace}\left\{\boldsymbol{C}\boldsymbol{A}^{i}\right\}. \tag{6.24}$$

Therefore, from (6.23), we derive

$$\operatorname{rank}\left\{\boldsymbol{C}\left[\boldsymbol{A}^{i}\ \boldsymbol{A}^{i-1}\boldsymbol{B}_{\mathcal{S}_{K-i+1}}\ \boldsymbol{A}^{i-2}\boldsymbol{B}_{\mathcal{S}_{K-i+2}}\ \cdots\ \boldsymbol{B}_{\mathcal{S}_K}\right]\right\} = p. \tag{6.25}$$

Hence, we arrive at

$$\operatorname{rank}\left\{\boldsymbol{C}\boldsymbol{A}^{i}\right\} + \operatorname{rank}\left\{\boldsymbol{C}\left[\boldsymbol{A}^{i-1}\boldsymbol{B}_{\mathcal{S}_{K-i+1}}\ \boldsymbol{A}^{i-2}\boldsymbol{B}_{\mathcal{S}_{K-i+2}}\ \cdots\ \boldsymbol{B}_{\mathcal{S}_K}\right]\right\} \ge p. \tag{6.26}$$

Also, since $\boldsymbol{C}\left[\boldsymbol{A}^{i-1}\boldsymbol{B}_{\mathcal{S}_{K-i+1}}\ \boldsymbol{A}^{i-2}\boldsymbol{B}_{\mathcal{S}_{K-i+2}}\ \cdots\ \boldsymbol{B}_{\mathcal{S}_K}\right] \in \mathbb{R}^{p\times is}$, we get

$$\operatorname{rank}\left\{\boldsymbol{C}\boldsymbol{A}^{i}\right\} + is \ge p. \tag{6.27}$$

Rewriting the above expression, we obtain for all $i = 1, 2 \ldots,$

$$s \ge \frac{1}{i}\left(p - \operatorname{rank}\left\{\boldsymbol{C}\boldsymbol{A}^{i}\right\}\right), \tag{6.28}$$

which leads to the desired necessary condition. □

Next, we provide sufficient conditions for sparse output controllability. We first derive the results for controllable linear dynamical systems, i.e., we assume that the controllability matrix $\boldsymbol{\Phi}$ in (6.21) has full rank, i.e., rank $\{\boldsymbol{\Phi}\} = n$. We relax later this controllability assumption in Theorem 6.5.

Proposition 6.2 *Suppose that the linear dynamical system* $(\boldsymbol{A} \in \mathbb{R}^{n\times n}, \boldsymbol{B} \in \mathbb{R}^{n\times m}, \boldsymbol{C} \in \mathbb{R}^{p\times n})$ *in* (6.18) *is controllable. Then, the system is s-sparse output controllable if*

$$\operatorname{rank}\{\boldsymbol{C}\boldsymbol{\Phi}\} = p \text{ and } s \ge \max_{1\le i\le n} \operatorname{rank}\left\{\boldsymbol{C}\boldsymbol{A}^{i}\right\} - \operatorname{rank}\left\{\boldsymbol{C}\boldsymbol{A}^{i-1}\right\}, \tag{6.29}$$

where $\boldsymbol{\Phi}$ *is the controllability matrix defined in* (6.21).

We isolate three lemmas to prove that the conditions are sufficient.

Lemma 6.1 *Suppose that the linear dynamical system* ($\boldsymbol{A} \in \mathbb{R}^{n\times n}$, $\boldsymbol{B} \in \mathbb{R}^{n\times m}$, $\boldsymbol{C} \in \mathbb{R}^{p\times n}$) *in* (6.18) *is controllable. Then, for* $0 \le i \le n-1$,

$$\text{ColSpace}\left\{\boldsymbol{R}^{(i)} - \boldsymbol{R}^{(i+1)}\right\} = \text{ColSpace}\left\{\left[\boldsymbol{I} - \boldsymbol{R}^{(i+1)}\right]\boldsymbol{C}\boldsymbol{A}^i\right\}, \tag{6.30}$$

where $\boldsymbol{R}^{(i)} = \boldsymbol{C}\boldsymbol{A}^i\left(\boldsymbol{C}\boldsymbol{A}^i\right)^\dagger$. *Also,*

$$\text{rank}\left\{\boldsymbol{R}^{(i)} - \boldsymbol{R}^{(i+1)}\right\} = \text{rank}\left\{\boldsymbol{C}\boldsymbol{A}^i\right\} - \text{rank}\left\{\boldsymbol{C}\boldsymbol{A}^{i+1}\right\}. \tag{6.31}$$

Proof We start by noting that for any $0 \le i \le n-1$, $\boldsymbol{R}^{(i)}$ and $\boldsymbol{R}^{(i+1)}$ are symmetric and

$$\boldsymbol{R}^{(i+1)} = \boldsymbol{R}^{(i+1)\mathsf{T}} = \left[\boldsymbol{C}\boldsymbol{A}^{i+1}\left(\boldsymbol{C}\boldsymbol{A}^{i+1}\right)^\dagger\right]^\mathsf{T} \tag{6.32}$$

$$= \left[\boldsymbol{C}\boldsymbol{A}^i\boldsymbol{A}\left(\boldsymbol{C}\boldsymbol{A}^{i+1}\right)^\dagger\right]^\mathsf{T} \tag{6.33}$$

$$= \left[\boldsymbol{C}\boldsymbol{A}^i\left(\boldsymbol{C}\boldsymbol{A}^i\right)^\dagger\boldsymbol{C}\boldsymbol{A}^i\boldsymbol{A}\left(\boldsymbol{C}\boldsymbol{A}^{i+1}\right)^\dagger\right]^\mathsf{T} \tag{6.34}$$

$$= \left[\boldsymbol{R}^{(i)}\boldsymbol{R}^{(i+1)}\right]^\mathsf{T} \tag{6.35}$$

$$= \boldsymbol{R}^{(i+1)}\boldsymbol{R}^{(i)}. \tag{6.36}$$

Therefore, we derive

$$\text{ColSpace}\left\{\boldsymbol{R}^{(i)} - \boldsymbol{R}^{(i+1)}\right\} = \text{ColSpace}\left\{\boldsymbol{R}^{(i)} - \boldsymbol{R}^{(i+1)}\boldsymbol{R}^{(i)}\right\} \tag{6.37}$$

$$= \text{ColSpace}\left\{\left[\boldsymbol{I} - \boldsymbol{R}^{(i+1)}\right]\boldsymbol{R}^{(i)}\right\} \tag{6.38}$$

$$= \text{ColSpace}\left\{\left[\boldsymbol{I} - \boldsymbol{R}^{(i+1)}\right]\boldsymbol{C}\boldsymbol{A}^i\right\}, \tag{6.39}$$

which follows because $\text{ColSpace}\left\{\boldsymbol{R}^{(i)}\right\} = \text{ColSpace}\left\{\boldsymbol{C}\boldsymbol{A}^i\right\}$.

Furthermore, since $\text{ColSpace}\left\{\boldsymbol{C}\boldsymbol{A}^{i+1}\right\} \subseteq \text{ColSpace}\left\{\boldsymbol{C}\boldsymbol{A}^i\right\}$, we also deduce

$$\text{rank}\left\{\boldsymbol{C}\boldsymbol{A}^i\right\} = \text{rank}\left\{\boldsymbol{C}\boldsymbol{A}^{i+1}\right\} + \text{rank}\left\{\left[\boldsymbol{I} - \boldsymbol{C}\boldsymbol{A}^{i+1}\left(\boldsymbol{C}\boldsymbol{A}^{i+1}\right)^\dagger\right]\boldsymbol{C}\boldsymbol{A}^i\right\} \tag{6.40}$$

$$= \text{rank}\left\{\boldsymbol{C}\boldsymbol{A}^{i+1}\right\} + \text{rank}\left\{\left[\boldsymbol{I} - \boldsymbol{R}^{(i+1)}\right]\boldsymbol{C}\boldsymbol{A}^i\right\} \tag{6.41}$$

$$= \text{rank}\left\{\boldsymbol{C}\boldsymbol{A}^{i+1}\right\} + \text{rank}\left\{\boldsymbol{R}^{(i)} - \boldsymbol{R}^{(i+1)}\right\}, \tag{6.42}$$

where we use (6.39). Thus, we have the rank condition. □

Lemma 6.2 *Suppose that the linear dynamical system* ($\boldsymbol{A} \in \mathbb{R}^{n\times n}$, $\boldsymbol{B} \in \mathbb{R}^{n\times m}$, $\boldsymbol{C} \in \mathbb{R}^{p\times n}$) *in* (6.18) *is controllable. Then, for* $0 \le i \le n-1$, *there exists a set* $\mathcal{S} \subseteq \{1, 2, \ldots, m\}$ *such that* $|\mathcal{S}| = \text{rank}\left\{\boldsymbol{C}\boldsymbol{A}^i\right\} - \text{rank}\left\{\boldsymbol{C}\boldsymbol{A}^{i+1}\right\}$ *and*

$$\text{ColSpace}\left\{\boldsymbol{R}^{(i)} - \boldsymbol{R}^{(i+1)}\right\} = \text{ColSpace}\left\{\left[\boldsymbol{I} - \boldsymbol{R}^{(i+1)}\right]\boldsymbol{C}\boldsymbol{A}^i\boldsymbol{B}_\mathcal{S}\right\}, \tag{6.43}$$

where $\boldsymbol{R}^{(i)} = \boldsymbol{C}\boldsymbol{A}^i\left(\boldsymbol{C}\boldsymbol{A}^i\right)^\dagger$.

Proof Since the system is controllable, from Theorem 4.1 and (6.21), we have ColSpace $\{\boldsymbol{\Phi}\} = \mathbb{R}^n$. Therefore, from Lemma 6.1, we have

$$\text{ColSpace}\left\{\boldsymbol{R}^{(i)} - \boldsymbol{R}^{(i+1)}\right\} = \text{ColSpace}\left\{\left[\boldsymbol{I} - \boldsymbol{R}^{(i+1)}\right]\boldsymbol{C}\boldsymbol{A}^i\boldsymbol{\Phi}\right\} \tag{6.44}$$

$$= \text{ColSpace}\left\{\left[\boldsymbol{I} - \boldsymbol{R}^{(i+1)}\right]\boldsymbol{C}\boldsymbol{A}^i\boldsymbol{B}\right\}, \tag{6.45}$$

which follows from the fact that, for any $k \geq 1$,

$$\left[\boldsymbol{I} - \boldsymbol{R}^{(i+1)}\right]\boldsymbol{C}\boldsymbol{A}^i\boldsymbol{A}^k = \boldsymbol{C}\boldsymbol{A}^{i+k} - \boldsymbol{C}\boldsymbol{A}^{i+1}\left(\boldsymbol{C}\boldsymbol{A}^{i+1}\right)^{\dagger}\boldsymbol{C}\boldsymbol{A}^{i+k} = \boldsymbol{0}. \tag{6.46}$$

Also, from Lemma 6.1, we obtain

$$\text{rank}\left\{\boldsymbol{R}^{(i)} - \boldsymbol{R}^{(i+1)}\right\} = \text{rank}\left\{\boldsymbol{C}\boldsymbol{A}^i\right\} - \text{rank}\left\{\boldsymbol{C}\boldsymbol{A}^{i+1}\right\} = |\mathcal{S}|. \tag{6.47}$$

Thus, from (6.45), there exist $|\mathcal{S}| = \text{rank}\left\{\boldsymbol{R}^{(i)} - \boldsymbol{R}^{(i+1)}\right\}$ columns in the matrix $\left[\boldsymbol{I} - \boldsymbol{R}^{(i+1)}\right]\boldsymbol{C}\boldsymbol{A}^i\boldsymbol{B}$ that span ColSpace $\left\{\boldsymbol{R}^{(i)} - \boldsymbol{R}^{(i+1)}\right\}$, and the proof is complete. □

Lemma 6.3 *Suppose that the linear dynamical system ($\boldsymbol{A} \in \mathbb{R}^{n\times n}$, $\boldsymbol{B} \in \mathbb{R}^{n\times m}$, $\boldsymbol{C} \in \mathbb{R}^{p\times n}$) in (6.18) is controllable. Then, for any sparsity level $s \leq m$, there exist an integer $K < \infty$ and sets $\{\mathcal{S}_k \subseteq \{1, 2, \ldots, m\} : |\mathcal{S}_k| = s\}_{k=1}^K$ such that*

$$\text{ColSpace}\left\{\boldsymbol{C}\boldsymbol{A}^n\right\} \subseteq \text{ColSpace}\left\{\boldsymbol{C}\boldsymbol{A}^n\left[\boldsymbol{A}^{K-1}\boldsymbol{B}_{\mathcal{S}_1}\ \boldsymbol{A}^{K-2}\boldsymbol{B}_{\mathcal{S}_2}\ \ldots\ \boldsymbol{B}_{\mathcal{S}_K}\right]\right\}. \tag{6.48}$$

Proof Let the real Jordan canonical of $\boldsymbol{A}$ be

$$\boldsymbol{A} = \boldsymbol{P}^{-1}\begin{bmatrix}\boldsymbol{J} & \boldsymbol{0}\\ \boldsymbol{0} & \boldsymbol{N}\end{bmatrix}\boldsymbol{P}, \tag{6.49}$$

where $\boldsymbol{P} \in \mathbb{R}^{n\times n}$ is an invertible matrix and the square matrices $\boldsymbol{J}$ and $\boldsymbol{N}$ are formed by the Jordan blocks of $\boldsymbol{A}$ corresponding to its nonzero and zero eigenvalues, respectively. Here, $\boldsymbol{J}$ is an invertible matrix, and $\boldsymbol{N}$ is a nilpotent matrix, i.e., $\boldsymbol{N}^n = \boldsymbol{0}$. Consequently, for any $n \leq P$, we deduce

$$\boldsymbol{A}^P = \boldsymbol{P}^{-1}\begin{bmatrix}\boldsymbol{J}^P & \boldsymbol{0}\\ \boldsymbol{0} & \boldsymbol{0}\end{bmatrix}\boldsymbol{P}. \tag{6.50}$$

Also, for any $n \leq P \leq Q$, by the Cayley–Hamilton theorem, we know that there exist real numbers $\{\alpha_j\}_{j=0}^{n-1}$ such that $\boldsymbol{J}^{P-Q} = \sum_{j=0}^{n-1}\alpha_j\boldsymbol{J}^j$. Substituting this relation into (6.50) yields

$$\boldsymbol{A}^P = \boldsymbol{P}^{-1}\begin{bmatrix}\boldsymbol{J}^Q\sum_{j=0}^{n-1}\alpha_j\boldsymbol{J}^j & \boldsymbol{0}\\ \boldsymbol{0} & \boldsymbol{0}\end{bmatrix}\boldsymbol{P} = \sum_{j=0}^{n-1}\alpha_j\boldsymbol{P}^{-1}\begin{bmatrix}\boldsymbol{J}^{Q+j} & \boldsymbol{0}\\ \boldsymbol{0} & \boldsymbol{0}\end{bmatrix}\boldsymbol{P} = \sum_{j=0}^{n-1}\alpha_j\boldsymbol{A}^{Q+j}. \tag{6.51}$$

Therefore, for any $n \leq P \leq Q$, and set $\mathcal{T} \subseteq \{1, 2, \ldots, m\}$, we arrive at

$$\text{ColSpace}\left\{\boldsymbol{A}^P \boldsymbol{B}_{\mathcal{T}}\right\} = \text{ColSpace}\left\{\sum_{j=0}^{n-1} \alpha_j \boldsymbol{A}^{Q+j} \boldsymbol{B}_{\mathcal{T}}\right\} \tag{6.52}$$

$$\subseteq \text{ColSpace}\left\{\boldsymbol{A}^Q \left[\boldsymbol{A}^{n-1} \boldsymbol{B}_{\mathcal{T}}\ \boldsymbol{A}^{n-2} \boldsymbol{B}_{\mathcal{T}}\ \ldots\ \boldsymbol{B}_{\mathcal{T}}\right]\right\}. \tag{6.53}$$

Hence, for any $\{\mathcal{T}_l\}_{l=1}^m$ such that $|\mathcal{T}_l| = s$ and $\cup_{l=1}^m \mathcal{T}_l = \{1, 2, \ldots, m\}$, we deduce

$$\text{ColSpace}\left\{\boldsymbol{A}^P \boldsymbol{B}\right\} = \text{ColSpace}\left\{\left[\boldsymbol{A}^P \boldsymbol{B}_{\mathcal{T}_1}\ \boldsymbol{A}^P \boldsymbol{B}_{\mathcal{T}_2}\ \ldots\ \boldsymbol{A}^P \boldsymbol{B}_{\mathcal{T}_m}\right]\right\}. \tag{6.54}$$

Here, substituting $\mathcal{T} = \mathcal{T}_l$ and $Q = (P - n)nm + n + n(m - l)$ in (6.53), we get

$$\text{ColSpace}\left\{\boldsymbol{A}^P \boldsymbol{B}_{\mathcal{T}_l}\right\} \subseteq$$
$$\text{ColSpace}\left\{\boldsymbol{A}^{(P-n)nm+n} \times \left[\boldsymbol{A}^{n(m-l)+n-1} \boldsymbol{B}_{\mathcal{T}_l}\ \boldsymbol{A}^{n(m-l)+n-2} \boldsymbol{B}_{\mathcal{T}_l}\ \ldots\ \boldsymbol{A}^{n(m-l)} \boldsymbol{B}_{\mathcal{T}_l}\right]\right\}. \tag{6.55}$$

Combining the above relation with (6.54) gives

$$\begin{aligned}
&\text{ColSpace}\left\{\boldsymbol{A}^P \boldsymbol{B}\right\} \\
&\quad \subseteq \text{ColSpace}\Big\{\boldsymbol{A}^{(P-n)nm+n}\big[\boldsymbol{A}^{nm-1} \boldsymbol{B}_{\mathcal{T}_1}\ \boldsymbol{A}^{nm-2} \boldsymbol{B}_{\mathcal{T}_1}\ \ldots\ \boldsymbol{A}^{n(m-1)} \boldsymbol{B}_{\mathcal{T}_1} \\
&\qquad \boldsymbol{A}^{n(m-1)-1} \boldsymbol{B}_{\mathcal{T}_2}\ \boldsymbol{A}^{n(m-1)-2} \boldsymbol{B}_{\mathcal{T}_2}\ \ldots\ \boldsymbol{A}^{n(m-2)} \boldsymbol{B}_{\mathcal{T}_2}\ \ldots \\
&\qquad\quad \boldsymbol{A}^{n-1} \boldsymbol{B}_{\mathcal{T}_m}\ \boldsymbol{A}^{n-2} \boldsymbol{B}_{\mathcal{T}_m}\ \ldots\ \boldsymbol{B}_{\mathcal{T}_m}\big]\Big\}.
\end{aligned} \tag{6.56}$$

Further, substituting for $P = n, n + 1, \ldots, 2n - 1$ in the above relation, we obtain that, for any sparsity level $s \leq m$, with $K = n^2 m$ and $\mathcal{S}_k = \mathcal{T}_{m-(\lfloor k/n \rfloor \bmod m)}$,

$$\begin{aligned}
&\text{ColSpace}\left\{\left[\boldsymbol{A}^{2n-1} \boldsymbol{B}\ \boldsymbol{A}^{2n-2} \boldsymbol{B}\ \ldots\ \boldsymbol{A}^n \boldsymbol{B}\right]\right\} \\
&\subseteq \text{ColSpace}\left\{\boldsymbol{A}^n \left[\boldsymbol{A}^{K-1} \boldsymbol{B}_{\mathcal{S}_1}\ \boldsymbol{A}^{K-2} \boldsymbol{B}_{\mathcal{S}_2}\ \ldots\ \boldsymbol{B}_{\mathcal{S}_K}\right]\right\}.
\end{aligned} \tag{6.57}$$

Finally, the above relation implies that

$$\text{ColSpace}\left\{\boldsymbol{C}\boldsymbol{A}^n \left[\boldsymbol{A}^{K-1} \boldsymbol{B}_{\mathcal{S}_1}\ \boldsymbol{A}^{K-2} \boldsymbol{B}_{\mathcal{S}_2}\ \ldots\ \boldsymbol{B}_{\mathcal{S}_K}\right]\right\}$$
$$\supseteq \text{ColSpace}\left\{\boldsymbol{C}\left[\boldsymbol{A}^{2n-1} \boldsymbol{B}\ \boldsymbol{A}^{2n-2} \boldsymbol{B}\ \ldots\ \boldsymbol{A}^n \boldsymbol{B}\right]\right\} \tag{6.58}$$
$$= \text{ColSpace}\left\{\boldsymbol{C}\boldsymbol{A}^n \left[\boldsymbol{A}^{n-1} \boldsymbol{B}\ \boldsymbol{A}^{n-2} \boldsymbol{B}\ \ldots\ \boldsymbol{B}\right]\right\} \tag{6.59}$$
$$= \text{ColSpace}\left\{\boldsymbol{C}\boldsymbol{A}^n\right\}, \tag{6.60}$$

where we use the fact that the system is controllable and

$$\text{ColSpace}\left\{\left[\boldsymbol{A}^{n-1} \boldsymbol{B}\ \boldsymbol{A}^{n-2} \boldsymbol{B}\ \ldots\ \boldsymbol{B}\right]\right\} = \mathbb{R}^n. \tag{6.61}$$

Hence, the proof is completed. □

We are now ready to prove the sufficient conditions for sparse output controllability.

Proof (of Proposition 6.2) Suppose the sufficient conditions in Proposition 6.2 hold. Then, for any $z \in \mathbb{R}^p$ and $i = 0, 1, \ldots, n-1$, we have

$$\boldsymbol{R}^{(i)} z = \left[\boldsymbol{R}^{(i)} - \boldsymbol{R}^{(i+1)}\right] z + \boldsymbol{R}^{(i+1)} z, \tag{6.62}$$

where $\boldsymbol{R}^{(i)} = \boldsymbol{C}\boldsymbol{A}^i \left(\boldsymbol{C}\boldsymbol{A}^i\right)^\dagger$. Also, using Lemma 6.2, there exists a vector $\boldsymbol{u} \in \mathbb{R}^m$ such that $\|\boldsymbol{u}\|_0 = \operatorname{rank}\left\{\boldsymbol{R}^{(i)}\right\} - \operatorname{rank}\left\{\boldsymbol{R}^{(i+1)}\right\} \le s$ such that

$$\boldsymbol{R}^{(i)} z = \left[\boldsymbol{I} - \boldsymbol{R}^{(i+1)}\right] \boldsymbol{C}\boldsymbol{B}\boldsymbol{u} + \boldsymbol{R}^{(i+1)} z = \boldsymbol{C}\boldsymbol{B}\boldsymbol{u} + \boldsymbol{R}^{(i+1)} \left[z - \boldsymbol{C}\boldsymbol{B}\boldsymbol{u}\right]. \tag{6.63}$$

Therefore, recursively using (6.63) with $z = \boldsymbol{y} - \sum_{k=1}^{l} \boldsymbol{C}\boldsymbol{A}^{k-1}\boldsymbol{B}\boldsymbol{u}_k$, we conclude that for any $\boldsymbol{y} \in \mathbb{R}^p$,

$$\boldsymbol{R}^{(0)} \boldsymbol{y} = \boldsymbol{C}\boldsymbol{B}\boldsymbol{u}_1 + \boldsymbol{R}^{(1)} \left[\boldsymbol{y} - \boldsymbol{C}\boldsymbol{B}\boldsymbol{u}_1\right] \tag{6.64}$$

$$= \boldsymbol{C}\boldsymbol{B}\boldsymbol{u}_1 + \boldsymbol{C}\boldsymbol{A}\boldsymbol{B}\boldsymbol{u}_2 + \boldsymbol{R}^{(2)} \left[\boldsymbol{y} - \boldsymbol{C}\boldsymbol{B}\boldsymbol{u}_1 - \boldsymbol{C}\boldsymbol{A}\boldsymbol{B}\boldsymbol{u}_2\right] \tag{6.65}$$

$$= \sum_{k=1}^{n} \boldsymbol{C}\boldsymbol{A}^{k-1}\boldsymbol{B}\boldsymbol{u}_k + \boldsymbol{R}^{(n)} \left[\boldsymbol{y} - \sum_{k=1}^{n} \boldsymbol{C}\boldsymbol{A}^{k-1}\boldsymbol{B}\boldsymbol{u}_k\right], \tag{6.66}$$

for some $\boldsymbol{u}_k$ such that $\|\boldsymbol{u}_k\|_0 \le s$, for $k = 1, 2, \ldots, n$. Here, we note that

$$p \ge \operatorname{rank}\{\boldsymbol{C}\} \ge \operatorname{rank}\{\boldsymbol{C}\boldsymbol{\Phi}\} = p. \tag{6.67}$$

So, we have $\boldsymbol{R}^{(0)} = \boldsymbol{C}\boldsymbol{C}^\dagger = \boldsymbol{I}$, and hence, for any $\boldsymbol{y} \in \mathbb{R}^p$, there exist s-sparse vectors $\{\boldsymbol{u}_k\}_{k=1}^n$ such that

$$\boldsymbol{y} = \sum_{k=1}^{n} \boldsymbol{C}\boldsymbol{A}^{k-1}\boldsymbol{B}\boldsymbol{u}_k + \boldsymbol{R}^{(n)} \left[\boldsymbol{y} - \sum_{k=1}^{n} \boldsymbol{C}\boldsymbol{A}^{k-1}\boldsymbol{B}\boldsymbol{u}_k\right]. \tag{6.68}$$

Here, the last term on the right-hand side belongs to ColSpace $\left\{\boldsymbol{C}\boldsymbol{A}^n\right\}$. Consequently, using Lemma 6.3, we derive that, for any $\boldsymbol{y} \in \mathbb{R}^p$, there exist an integer $K < \infty$ and s-sparse vectors $\{\boldsymbol{u}_k\}_{k=1}^K$ such that

$$\boldsymbol{y} = \sum_{k=1}^{K} \boldsymbol{C}\boldsymbol{A}^{k-1}\boldsymbol{B}\boldsymbol{u}_k. \tag{6.69}$$

Hence, from Theorem 6.4, we conclude that the linear dynamical system $(\boldsymbol{A}, \boldsymbol{B}, \boldsymbol{C})$ is s-sparse output controllable. □

As promised, we now relax the assumption that the system is controllable to present the most general result for sparse output controllability.

Theorem 6.5 *The linear dynamical system* ($\boldsymbol{A} \in \mathbb{R}^{n\times n}$, $\boldsymbol{B} \in \mathbb{R}^{n\times m}$, $\boldsymbol{C} \in \mathbb{R}^{p\times n}$) *in* (6.18) *is s-sparse output controllable only if*

$$\operatorname{rank}\{\boldsymbol{C\Phi}\} = p \text{ and } s \geq \max_{1\leq i\leq n} \frac{1}{i}\left(p - \operatorname{rank}\left\{\boldsymbol{C}\boldsymbol{A}^i\boldsymbol{\Phi}\right\}\right), \tag{6.70}$$

where $\boldsymbol{\Phi}$ is the controllability matrix defined in (6.21)*. The system is s-sparse output controllable if*

$$\operatorname{rank}\{\boldsymbol{C\Phi}\} = p \text{ and } s \geq \max_{0\leq i\leq n-1} \operatorname{rank}\left\{\boldsymbol{C}\boldsymbol{A}^i\boldsymbol{\Phi}\right\} - \operatorname{rank}\left\{\boldsymbol{C}\boldsymbol{A}^{i+1}\boldsymbol{\Phi}\right\}. \tag{6.71}$$

The necessary conditions in (6.70) directly follows from Proposition 6.1. So, it suffices to prove the sufficient conditions in (6.71). To this end, let $\boldsymbol{Q} \in \mathbb{R}^{n\times R}$ be a matrix whose columns are orthonormal and ColSpace $\{\boldsymbol{Q}\}$ = ColSpace $\{\boldsymbol{\Phi}\}$ with $R = \operatorname{rank}\{\boldsymbol{\Phi}\}$. Further, the Kalman decomposition ensures that there exists an orthogonal matrix $\bar{\boldsymbol{Q}} = \left[\boldsymbol{Q}\ \boldsymbol{\Phi}^{\perp}\right] \in \mathbb{R}^{n\times n}$ such that

$$\bar{\boldsymbol{Q}}^{\mathsf{T}}\boldsymbol{A}\bar{\boldsymbol{Q}} = \begin{bmatrix} \tilde{\boldsymbol{A}} & \boldsymbol{Q}^{\mathsf{T}}\boldsymbol{A}\boldsymbol{\Phi}^{\perp} \\ \boldsymbol{0} & (\boldsymbol{\Phi}^{\perp})^{\mathsf{T}}\boldsymbol{A}\boldsymbol{\Phi}^{\perp} \end{bmatrix} \text{ and } \bar{\boldsymbol{Q}}^{\mathsf{T}}\boldsymbol{B} = \begin{bmatrix} \tilde{\boldsymbol{B}} \\ \boldsymbol{0} \end{bmatrix}, \tag{6.72}$$

where $\tilde{\boldsymbol{A}} = \boldsymbol{Q}^{\mathsf{T}}\boldsymbol{A}\boldsymbol{Q}$ and $\tilde{\boldsymbol{B}} = \boldsymbol{Q}^{\mathsf{T}}\boldsymbol{B}$. Consequently, for any integer $i \geq 0$, we have

$$\begin{aligned} \boldsymbol{C}\boldsymbol{A}^i\boldsymbol{B} &= \boldsymbol{C}\bar{\boldsymbol{Q}}\bar{\boldsymbol{Q}}^T\boldsymbol{A}^i\bar{\boldsymbol{Q}}\bar{\boldsymbol{Q}}^T\boldsymbol{B} \\ &= \boldsymbol{C}\left[\boldsymbol{Q}\ \boldsymbol{\Phi}^{\perp}\right]\begin{bmatrix} \tilde{\boldsymbol{A}}^i\tilde{\boldsymbol{B}} \\ \boldsymbol{0} \end{bmatrix} = \begin{bmatrix} (\boldsymbol{C}\boldsymbol{Q})\tilde{\boldsymbol{A}}^i\tilde{\boldsymbol{B}} \\ \boldsymbol{0} \end{bmatrix}. \end{aligned} \tag{6.73}$$

This formulation leads to a modified system ($\tilde{\boldsymbol{A}} \in \mathbb{R}^{R\times R}$, $\tilde{\boldsymbol{B}} \in \mathbb{R}^{R\times m}$, $\boldsymbol{C}\boldsymbol{Q} \in \mathbb{R}^{p\times R}$) with the controllability matrix

$$\tilde{\boldsymbol{\Phi}} = \left[\tilde{\boldsymbol{A}}^{K-1}\tilde{\boldsymbol{B}}\ \tilde{\boldsymbol{A}}^{K-2}\tilde{\boldsymbol{B}} \ldots \tilde{\boldsymbol{B}}\right]. \tag{6.74}$$

From (6.73) and 6.4, this modified system is s-sparse output controllable if and only if there exist an integer $K < \infty$ and sets $\{\mathcal{S}_k \subseteq \{1, 2, \ldots, m\},\ |\mathcal{S}_k| \leq s\}_{k=1}^K$ such that

$$\begin{aligned} &\operatorname{rank}\left\{\boldsymbol{C}\boldsymbol{Q}\left[\tilde{\boldsymbol{A}}^{K-1}\tilde{\boldsymbol{B}}_{\mathcal{S}_1}\ \tilde{\boldsymbol{A}}^{K-2}\tilde{\boldsymbol{B}}_{\mathcal{S}_2} \ldots \tilde{\boldsymbol{B}}_{\mathcal{S}_K}\right]\right\} \\ &\qquad = \operatorname{rank}\left\{\boldsymbol{C}\left[\boldsymbol{A}^{K-1}\boldsymbol{B}_{\mathcal{S}_1}\ \boldsymbol{A}^{K-2}\boldsymbol{B}_{\mathcal{S}_2} \ldots \boldsymbol{B}_{\mathcal{S}_K}\right]\right\} = p. \end{aligned} \tag{6.75}$$

So, we see that the modified system is s-sparse output controllable if and only if the system $(\boldsymbol{A}, \boldsymbol{B}, \boldsymbol{C})$ is s-sparse output controllable. It is also easy to see that

$\text{rank}\left\{\tilde{\boldsymbol{\Phi}}\right\} = \text{rank}\left\{\boldsymbol{\Phi}\right\} = R$, and as a result, from Proposition 6.2, the modified system is s-sparse output controllable if

$$\text{rank}\left\{\boldsymbol{C}\boldsymbol{Q}\tilde{\boldsymbol{\Phi}}\right\} = p \quad \text{and} \quad \text{rank}\left\{(\boldsymbol{C}\boldsymbol{Q})\,\tilde{\boldsymbol{A}}^{i}\right\} - \text{rank}\left\{(\boldsymbol{C}\boldsymbol{Q})\,\tilde{\boldsymbol{A}}^{i+1}\right\} \leq s. \tag{6.76}$$

Therefore, to prove the sufficient conditions, it is enough to verify that (6.76) is satisfied when (6.71) holds.

Furthermore, using (6.73) and assuming (6.71) holds, the controllability matrix of the modified system satisfies

$$\begin{aligned}\text{rank}\left\{\boldsymbol{C}\boldsymbol{Q}\tilde{\boldsymbol{\Phi}}\right\} &= \text{rank}\left\{\boldsymbol{C}\boldsymbol{Q}\left[\tilde{\boldsymbol{A}}^{K-1}\tilde{\boldsymbol{B}}\;\tilde{\boldsymbol{A}}^{K-2}\tilde{\boldsymbol{B}}\ldots\tilde{\boldsymbol{B}}\right]\right\}\\ &= \text{rank}\left\{\boldsymbol{C}\left[\boldsymbol{A}^{K-1}\boldsymbol{B}\;\boldsymbol{A}^{K-2}\boldsymbol{B}\ldots\boldsymbol{B}\right]\right\} = \text{rank}\left\{\boldsymbol{C}\boldsymbol{\Phi}\right\} = p.\end{aligned} \tag{6.77}$$

Similarly, (6.72), using we have

$$\begin{aligned}\text{rank}\left\{(\boldsymbol{C}\boldsymbol{Q})\,\tilde{\boldsymbol{A}}^{i}\right\} &= \text{rank}\left\{\left[\boldsymbol{C}\boldsymbol{Q}\;\;\boldsymbol{C}\boldsymbol{\Phi}^{\perp}\right]\begin{bmatrix}\tilde{\boldsymbol{A}} & \boldsymbol{Q}^{\mathsf{T}}\boldsymbol{A}\boldsymbol{\Phi}^{\perp}\\ \boldsymbol{0} & (\boldsymbol{\Phi}^{\perp})^{\mathsf{T}}\boldsymbol{A}\boldsymbol{\Phi}^{\perp}\end{bmatrix}^{i}\begin{bmatrix}\boldsymbol{I}\\ \boldsymbol{0}\end{bmatrix}\right\}\\ &= \text{rank}\left\{\boldsymbol{C}\bar{\boldsymbol{Q}}(\bar{\boldsymbol{Q}}^{\mathsf{T}}\boldsymbol{A}\bar{\boldsymbol{Q}})^{i}\begin{bmatrix}\boldsymbol{I}\\ \boldsymbol{0}\end{bmatrix}\right\} = \text{rank}\left\{\boldsymbol{C}\boldsymbol{A}^{i}\boldsymbol{Q}\right\} = \text{rank}\left\{\boldsymbol{C}\boldsymbol{A}^{i}\boldsymbol{\Phi}\right\}.\end{aligned} \tag{6.78}$$

Therefore, (6.71) implies (6.76), and the proof is complete.

Some observations from Theorem 6.5 are as follows. In Theorem 6.5, when $\boldsymbol{C} = \boldsymbol{I}$, the condition $\text{rank}\left\{\boldsymbol{C}\boldsymbol{\Phi}\right\} = p = n$ reduces to $\text{rank}\left\{\boldsymbol{\Phi}\right\} = n$. Also, the sufficient bound on the sparsity level becomes

$$\max_{0\leq i\leq n-1}\text{rank}\left\{\boldsymbol{C}\boldsymbol{A}^{i}\boldsymbol{\Phi}\right\} - \text{rank}\left\{\boldsymbol{C}\boldsymbol{A}^{i+1}\boldsymbol{\Phi}\right\} = \max_{0\leq i\leq n-1}\text{rank}\left\{\boldsymbol{A}^{i}\boldsymbol{\Phi}\right\} - \text{rank}\left\{\boldsymbol{A}^{i+1}\boldsymbol{\Phi}\right\} \tag{6.79}$$

$$= \max_{0\leq i\leq n-1}\text{rank}\left\{\boldsymbol{A}^{i}\right\} - \text{rank}\left\{\boldsymbol{A}^{i+1}\right\} \tag{6.80}$$

$$\leq n - \text{rank}\left\{\boldsymbol{A}\right\} \tag{6.81}$$

$$= \text{rank}\left\{\boldsymbol{C}\boldsymbol{\Phi}\right\} - \text{rank}\left\{\boldsymbol{C}\boldsymbol{A}\boldsymbol{\Phi}\right\}, \tag{6.82}$$

where we use the Sylvester rank inequality (see Theorem A.1) in (6.81). Therefore, we conclude that

$$\max_{0\leq i\leq n-1}\text{rank}\left\{\boldsymbol{C}\boldsymbol{A}^{i}\boldsymbol{\Phi}\right\} - \text{rank}\left\{\boldsymbol{C}\boldsymbol{A}^{i+1}\boldsymbol{\Phi}\right\} = n - \text{rank}\left\{\boldsymbol{A}\right\}. \tag{6.83}$$

So, the bound on the sparsity level is equivalent to $s \geq n - \text{rank}\left\{\boldsymbol{A}\right\}$. Similarly, the necessary bound on the sparsity level is

$$\max_{1\le i\le n}\frac{1}{i}\left(p-\operatorname{rank}\left\{\boldsymbol{C}\boldsymbol{A}^i\boldsymbol{\Phi}\right\}\right)=\max_{1\le i\le n}\frac{1}{i}\left(\operatorname{rank}\left\{\boldsymbol{A}^0\right\}-\operatorname{rank}\left\{\boldsymbol{A}^i\right\}\right) \tag{6.84}$$

$$=\max_{1\le i\le n}\frac{1}{i}\sum_{j=0}^{i-1}\operatorname{rank}\left\{\boldsymbol{A}^j\right\}-\operatorname{rank}\left\{\boldsymbol{A}^{j+1}\right\} \tag{6.85}$$

$$\le\max_{1\le i\le n}\operatorname{rank}\left\{\boldsymbol{A}^{i-1}\right\}-\operatorname{rank}\left\{\boldsymbol{A}^{i+1}\right\} \tag{6.86}$$

$$\le n-\operatorname{rank}\{\boldsymbol{A}\} \tag{6.87}$$

$$=n-\operatorname{rank}\{\boldsymbol{C}\boldsymbol{A}\boldsymbol{\Phi}\}. \tag{6.88}$$

Hence, the necessary bound in Theorem 6.5 reduces to $s\ge n-\operatorname{rank}\{\boldsymbol{A}\}$. Consequently, Theorem 6.5 reduces to Theorem 4.6 when $\boldsymbol{C}=\boldsymbol{I}$, implying Theorem 6.5 is a tight result.

Finally, the s-sparse control inputs required to drive the system to a desired output starting from a given initial state can be cast as a sparse signal recovery problem, similar to the ones discussed in Chapters 2 and 5.

6.3 Sparse Nonnegative Controllability

The linear dynamical system $(\boldsymbol{A},\boldsymbol{B})$ in (6.1) is said to be (non-sparse) nonnegative controllable if, for any initial state $\boldsymbol{x}_0$ and any final state $\boldsymbol{x}_{\mathrm{f}}$, there exist a time index $K<\infty$ and inputs $\{\boldsymbol{u}_k\in\mathbb{R}_+^m,\ k=1,2,\ldots,K\}$ that steer the system from the state $\boldsymbol{x}_0$ to $\boldsymbol{x}_K=\boldsymbol{x}_{\mathrm{f}}$. The classical test for nonnegative controllability is as follows:

Theorem 6.6 *The linear dynamical system* $(\boldsymbol{A}\in\mathbb{R}^{n\times n},\boldsymbol{B}\in\mathbb{R}^{n\times m})$ *in* (6.1) *is nonnegative controllable if and only if the following conditions hold:*

1. $\operatorname{rank}\left\{\left[\boldsymbol{A}-\lambda\boldsymbol{I}\ \boldsymbol{B}\right]\right\}=n$, *for all* $\lambda\in\mathbb{C}$
2. *For every left eigenvalue* $\lambda\ge 0$ *of* $\boldsymbol{A}$*, the corresponding eigenvector* z *is such that* $z^{\mathsf{T}}\boldsymbol{B}$ *has at least one strictly negative entry, i.e., there exist no tuples* $(\lambda\ge 0, z\ne\boldsymbol{0})$ *such* $z^{\mathsf{T}}\boldsymbol{A}=\lambda z^{\mathsf{T}}$ *and* $z^{\mathsf{T}}\boldsymbol{B}\in\mathbb{R}_+^m$.

We extend the notion of nonnegative controllability to the case of sparse inputs and derive a nonnegative controllability test as follows.

Definition 6.3 The linear dynamical system $(\boldsymbol{A},\boldsymbol{B})$ in (6.1) is said to be s-sparse nonnegative controllable for a given $s\ge 1$ if, for any initial state $\boldsymbol{x}_0$ and final state $\boldsymbol{x}_{\mathrm{f}}$, there exist an integer $K<\infty$ and s-sparse inputs $\{\boldsymbol{u}_k\in\mathbb{R}_+^m:\ \|\boldsymbol{u}_k\|_0\le s\}_{k=1}^K$ that steer the system from $\boldsymbol{x}_0$ to $\boldsymbol{x}_K=\boldsymbol{x}_{\mathrm{f}}$.

We note that sparse nonnegative controllability is a stronger notion than sparse controllability and nonnegative controllability. So, the conditions for s-sparse controllability in Theorem 4.6 and the conditions for nonnegative controllability in Theorem 6.6 are necessary for sparse nonnegative controllability. The main result of this section shows that the two sets of conditions are also sufficient for sparse nonnegative controllability.

Before presenting the main result, we look at a special case where $\boldsymbol{A}$ is invertible.

Proposition 6.3 *For any sparsity level, $s \geq 1$, the linear dynamical system ($\boldsymbol{A} \in \mathbb{R}^{n\times n}$, $\boldsymbol{B} \in \mathbb{R}^{n\times m}$) in (6.1), with nonsingular $\boldsymbol{A}$, is s-sparse nonnegative controllable if it is nonnegative controllable.*

Proof For a given sparsity level, suppose that the system is not s-sparse nonnegative controllable. Then, for any set $\mathcal{S} \subseteq \{1, 2, \ldots, m\}$ and $|\mathcal{S}| \leq s$, we can construct a set $\{\mathcal{T}_k \subseteq \{1, 2, \ldots, m\} : |\mathcal{T}_k| = s\}_{k=1}^{m-1}$ such that

$$\bigcup_{k=1}^{m-1} \mathcal{T}_k \bigcup \mathcal{S} = \{1, 2 \ldots, m\}. \tag{6.89}$$

Then, for any $K < \infty$, we have

$$\begin{aligned} \text{Span}_+ \Big\{ \big[& \boldsymbol{A}^{Km-1} \boldsymbol{B}_{\mathcal{T}_1} \; \boldsymbol{A}^{Km-2} \boldsymbol{B}_{\mathcal{T}_1} \; \ldots \; \boldsymbol{A}^{K(m-1)} \boldsymbol{B}_{\mathcal{T}_1} \\ & \boldsymbol{A}^{K(m-1)-1} \boldsymbol{B}_{\mathcal{T}_2} \; \boldsymbol{A}^{K(m-1)-2} \boldsymbol{B}_{\mathcal{T}_2} \; \ldots \; \boldsymbol{A}^{K(m-2)} \boldsymbol{B}_{\mathcal{T}_2} \\ & \ldots \; \boldsymbol{A}^{2K-1} \boldsymbol{B}_{\mathcal{T}_{m-1}} \; \boldsymbol{A}^{2K-2} \boldsymbol{B}_{\mathcal{T}_{m-1}} \; \ldots \; \boldsymbol{A}^{K} \boldsymbol{B}_{\mathcal{T}_{m-1}} \\ & \boldsymbol{A}^{K-1} \boldsymbol{B}_{\mathcal{S}} \; \boldsymbol{A}^{K-2} \boldsymbol{B}_{\mathcal{S}} \; \ldots \; \boldsymbol{B}_{\mathcal{S}} \big] \Big\} \subset \mathbb{R}^n, \end{aligned} \tag{6.90}$$

where Span_+ is the positive span of the columns of the matrix, i.e.,

$$\text{Span}_+ \{\boldsymbol{M}\} = \left\{ \sum_{i=1}^{M} \alpha_i \boldsymbol{M}_i : \; \alpha_i \geq 0 \right\}, \tag{6.91}$$

where M is the number of columns of $\boldsymbol{M}$. Rearranging the columns of the matrix, from (6.90), we derive

$$\begin{aligned} \text{Span}_+ \Big\{ \big[& \boldsymbol{A}^{Km-1} \boldsymbol{B}_{\mathcal{T}_1} \; \boldsymbol{A}^{K(m-1)-1} \boldsymbol{B}_{\mathcal{T}_2} \; \ldots \; \boldsymbol{A}^{2K-1} \boldsymbol{B}_{\mathcal{T}_{m-1}} \; \boldsymbol{A}^{K-1} \boldsymbol{B}_{\mathcal{S}} \\ & \boldsymbol{A}^{Km-2} \boldsymbol{B}_{\mathcal{T}_1} \; \boldsymbol{A}^{K(m-1)-2} \boldsymbol{B}_{\mathcal{T}_2} \; \ldots \; \boldsymbol{A}^{2K-2} \boldsymbol{B}_{\mathcal{T}_{m-1}} \; \boldsymbol{A}^{K-2} \boldsymbol{B}_{\mathcal{S}} \\ \ldots \; & \boldsymbol{A}^{K(m-1)} \boldsymbol{B}_{\mathcal{T}_1} \; \boldsymbol{A}^{K(m-2)} \boldsymbol{B}_{\mathcal{T}_2} \; \ldots \; \boldsymbol{A}^{K} \boldsymbol{B}_{\mathcal{T}_{m-1}} \; \boldsymbol{B}_{\mathcal{S}} \big] \Big\} \subset \mathbb{R}^n. \end{aligned} \tag{6.92}$$

We can further rewrite the above equation as

$$\text{Span}_+ \left\{ \left[\boldsymbol{A}^{K-1} \hat{\boldsymbol{B}} \; \boldsymbol{A}^{K-2} \hat{\boldsymbol{B}} \; \ldots \; \hat{\boldsymbol{B}} \right] \right\} \subset \mathbb{R}^n, \tag{6.93}$$

where we define

$$\hat{\boldsymbol{B}} = \left[\boldsymbol{A}^{K(m-1)} \boldsymbol{B}_{\mathcal{T}_1} \; \boldsymbol{A}^{K(m-2)} \boldsymbol{B}_{\mathcal{T}_2} \; \ldots \; \boldsymbol{A}^{K} \boldsymbol{B}_{\mathcal{T}_{m-1}} \; \boldsymbol{B}_{\mathcal{S}} \right]. \tag{6.94}$$

Since (6.93) holds for any $K < \infty$, we conclude that the system $(\boldsymbol{A}, \hat{\boldsymbol{B}})$ is not negative controllable. By Theorem 6.6, at least one of the following statements is violated:

1. There exists no tuple $(\lambda \in \mathbb{C}, \boldsymbol{z} \neq \boldsymbol{0})$ such that $\boldsymbol{z}^\mathsf{T}\boldsymbol{A} = \lambda\boldsymbol{z}^\mathsf{T}$ and $\boldsymbol{z}^\mathsf{T}\hat{\boldsymbol{B}} = \boldsymbol{0}$.
2. There exists no tuple $(\lambda \geq 0, \boldsymbol{z} \neq \boldsymbol{0})$ such that $\boldsymbol{z}^\mathsf{T}\boldsymbol{A} = \lambda\boldsymbol{z}^\mathsf{T}$ and $\boldsymbol{z}^\mathsf{T}\hat{\boldsymbol{B}} \in \mathbb{R}_+^m$.

Nonetheless, when $\boldsymbol{z}^\mathsf{T}\boldsymbol{A} = \lambda\boldsymbol{z}^\mathsf{T}$, we get

$$\boldsymbol{z}^\mathsf{T}\hat{\boldsymbol{B}} = \left[\lambda^{K(m-1)}\boldsymbol{z}^\mathsf{T}\boldsymbol{B}_{\mathcal{T}_1}\ \lambda^{K(m-2)}\boldsymbol{z}^\mathsf{T}\boldsymbol{B}_{\mathcal{T}_2}\ \ldots\ \lambda^{K}\boldsymbol{z}^\mathsf{T}\boldsymbol{B}_{\mathcal{T}_{m-1}}\ \boldsymbol{z}^\mathsf{T}\boldsymbol{B}_{\mathcal{S}}\right] \tag{6.95}$$

$$= \boldsymbol{z}^\mathsf{T}\left[\boldsymbol{B}_{\mathcal{T}_1}\ \boldsymbol{B}_{\mathcal{T}_2}\ \ldots\ \boldsymbol{B}_{\mathcal{T}_{m-1}}\ \boldsymbol{B}_{\mathcal{S}}\right]\begin{bmatrix}\lambda^{K(m-1)}\boldsymbol{I} & & & \\ & \lambda^{K(m-2)}\boldsymbol{I} & & \\ & & \ddots & \\ & & & \boldsymbol{I}\end{bmatrix}. \tag{6.96}$$

Also, since $\boldsymbol{A}$ is nonsingular, $\lambda \neq 0$, and as a consequence, using (6.89), at least one of the following statements are violated:

1. There exists no tuple $(\lambda \in \mathbb{C}, \boldsymbol{z} \neq \boldsymbol{0})$ such that $\boldsymbol{z}^\mathsf{T}\boldsymbol{A} = \lambda\boldsymbol{z}^\mathsf{T}$ and $\boldsymbol{z}^\mathsf{T}\boldsymbol{B} = \boldsymbol{0}$.
2. There exists no tuple $(\lambda \geq 0, \boldsymbol{z} \neq \boldsymbol{0})$ such that $\boldsymbol{z}^\mathsf{T}\boldsymbol{A} = \lambda\boldsymbol{z}^\mathsf{T}$ and $\boldsymbol{z}^\mathsf{T}\boldsymbol{B} \in \mathbb{R}_+^m$.

Therefore, by Theorem 6.6, the system $(\boldsymbol{A}, \boldsymbol{B})$ is not nonnegative controllable.

In short, we showed that if the system is not s-sparse nonnegative controllable, then it is not nonnegative controllable. Hence, the proof. □

Building on the above intermediate result, we present the main result of this section.

Theorem 6.7 *For any sparsity level, $s \geq 1$, the linear dynamical system $(\boldsymbol{A} \in \mathbb{R}^{n\times n}, \boldsymbol{B} \in \mathbb{R}^{n\times m})$ in* (6.1) *is s-sparse nonnegative controllable if and only if it is nonnegative controllable and s-sparse controllable.*

We prove the main result for the general system using the following two lemmas.

Lemma 6.4 *Suppose that the linear dynamical system $(\boldsymbol{A} \in \mathbb{R}^{n\times n}, \boldsymbol{B} \in \mathbb{R}^{n\times m})$ in* (6.1) *is nonnegative controllable and $s \geq n - \operatorname{rank}\{\boldsymbol{A}\}$. Then, for any $\boldsymbol{z} \in \mathbb{R}^n$ and integer $i = 0, 1, \ldots, n-1$, there exists an s-sparse vector $\boldsymbol{u} \in \mathbb{R}_+^m$ such that*

$$\left[\boldsymbol{A}^i\left(\boldsymbol{A}^i\right)^\dagger - \boldsymbol{A}^{i+1}\left(\boldsymbol{A}^{i+1}\right)^\dagger\right]\boldsymbol{z} = \left[\boldsymbol{I} - \boldsymbol{A}^{i+1}\left(\boldsymbol{A}^{i+1}\right)^\dagger\right]\boldsymbol{A}^i\boldsymbol{B}\boldsymbol{u}. \tag{6.97}$$

Proof Since the system is nonnegative controllable, for any $\boldsymbol{z} \in \mathbb{R}^n$, there exist an integer $K < \infty$ and (non-sparse) inputs $\{\boldsymbol{u}_k\}_{k=1}^K \in \mathbb{R}_+^m$ such that

$$\boldsymbol{z} = \sum_{k=1}^{K}\boldsymbol{A}^{K-k}\boldsymbol{B}\boldsymbol{u}_k. \tag{6.98}$$

Multiplying both sides by $\left[\boldsymbol{I} - \boldsymbol{A}^{i+1}\left(\boldsymbol{A}^{i+1}\right)^\dagger\right]\boldsymbol{A}^i$ yields

$$\left[I - A^{i+1}\left(A^{i+1}\right)^{\dagger}\right] A^{i} z = \sum_{k=1}^{K} \left[I - A^{i+1}\left(A^{i+1}\right)^{\dagger}\right] A^{i} A^{K-k} B u_{k} \tag{6.99}$$

$$= \left[I - A^{i+1}\left(A^{i+1}\right)^{\dagger}\right] A^{i} B u_{K}, \tag{6.100}$$

using the fact that, for $K - k \geq 1$,

$$\left[I - A^{i+1}\left(A^{i+1}\right)^{\dagger}\right] A^{K-k+i} = \left[A^{i+1} - A^{i+1}\left(A^{i+1}\right)^{\dagger} A^{i+1}\right] A^{K-k-1} = \mathbf{0}. \tag{6.101}$$

Also, from Lemma 6.1 with $C = I$, we know that

$$\text{ColSpace}\left\{\left[I - A^{i+1}\left(A^{i+1}\right)^{\dagger}\right] A^{i}\right\} = \text{ColSpace}\left\{A^{i}\left(A^{i}\right)^{\dagger} - A^{i+1}\left(A^{i+1}\right)^{\dagger}\right\}. \tag{6.102}$$

As a result, from (6.100), for every $z \in \mathbb{R}^{n}$, there exists a vector $v \in \mathbb{R}^{m}_{+}$ such that

$$\left[A^{i}\left(A^{i}\right)^{\dagger} - A^{i+1}\left(A^{i+1}\right)^{\dagger}\right] z = \left[I - A^{i+1}\left(A^{i+1}\right)^{\dagger}\right] A^{i} B v. \tag{6.103}$$

Using this formulation, we define a non-empty set

$$\mathcal{V}(z) = \left\{v \in \mathbb{R}^{n}_{+} : \left[A^{i}\left(A^{i}\right)^{\dagger} - A^{i+1}\left(A^{i+1}\right)^{\dagger}\right] z = \left[I - A^{i+1}\left(A^{i+1}\right)^{\dagger}\right] A^{i} B v\right\} \tag{6.104}$$

$$= \left\{v \in \mathbb{R}^{n}_{+} : \bar{z} = \bar{Z} v\right\}, \tag{6.105}$$

where $\bar{z} = \bar{Z} v$ is the row-reduced echelon form of the system of equations with variable v in (6.103) after removing the zero rows.

By the fundamental theorem in linear programming, the system $\bar{z} = \bar{Z} v$ has a basic feasible solution $v \in \mathbb{R}^{n}_{+}$ with at most rank $\left\{\bar{Z}\right\}$ nonzero elements. However, invoking Lemma 6.1 with $C = I$,

$$\text{rank}\left\{\bar{Z}\right\} = \text{rank}\left\{A^{i}\left(A^{i}\right)^{\dagger} - A^{i+1}\left(A^{i+1}\right)^{\dagger}\right\} = \text{rank}\left\{A^{i}\right\} - \text{rank}\left\{A^{i+1}\right\} \tag{6.106}$$

$$\leq n - \text{rank}\left\{A\right\} \leq s, \tag{6.107}$$

where we use the Sylvester rank inequality (see Theorem A.1). Hence, there exists an s-sparse vector $u \in \mathbb{R}^{m}_{+}$ such that

$$\left[A^{i}\left(A^{i}\right)^{\dagger} - A^{i+1}\left(A^{i+1}\right)^{\dagger}\right] z = \left[I - A^{i+1}\left(A^{i+1}\right)^{\dagger}\right] A^{i} B u. \tag{6.108}$$

The proof is complete. □

Lemma 6.5 *Consider the linear dynamical system* ($A \in \mathbb{R}^{n \times n}$, $B \in \mathbb{R}^{n \times m}$) *in* (6.1), *and any sparsity level* $s \leq m$. *Then, for any* $z \in \text{ColSpace}\left\{CA^{n}\right\}$, *there exist an* *integer $K < \infty$ and s-sparse inputs $\{\boldsymbol{u}_k \in \mathbb{R}_+^m\}_{k=1}^K$ such that*

$$z = \sum_{k=1}^{K} \boldsymbol{A}^{n+K-k} \boldsymbol{B} \boldsymbol{u}_k. \tag{6.109}$$

Proof Let the real Jordan canonical of $\boldsymbol{A}$ be

$$\boldsymbol{A} = \boldsymbol{P}^{-1} \begin{bmatrix} \boldsymbol{J} & \boldsymbol{0} \\ \boldsymbol{0} & \boldsymbol{N} \end{bmatrix} \boldsymbol{P}, \tag{6.110}$$

where $\boldsymbol{P} \in \mathbb{R}^{n\times n}$ is an invertible matrix and the square matrices $\boldsymbol{J} \in \mathbb{R}^{J\times J}$ and $\boldsymbol{N} \in \mathbb{R}^{(n-J)\times(n-J)}$ are formed by the Jordan blocks of $\boldsymbol{A}$ corresponding to its nonzero and zero eigenvalues, respectively. Here, $\boldsymbol{J}$ is an invertible matrix, and $\boldsymbol{N}$ is a nilpotent matrix, i.e., $\boldsymbol{N}^n = \boldsymbol{0}$. Consequently, for any $P \geq n$, we deduce

$$\boldsymbol{A}^P = \boldsymbol{P}^{-1} \begin{bmatrix} \boldsymbol{J}^P & \boldsymbol{0} \\ \boldsymbol{0} & \boldsymbol{0} \end{bmatrix} \boldsymbol{P}. \tag{6.111}$$

Also, due to the assumption that the system $(\boldsymbol{A}, \boldsymbol{B})$ is nonnegative controllable, we deduce that there exist an integer $K < \infty$ and (non-sparse) inputs $\{\boldsymbol{v}_k \in \mathbb{R}^m\}_{k=1}^K$ for any vector $\boldsymbol{x}$ such that

$$\boldsymbol{x} = \sum_{k=1}^{K} \boldsymbol{A}^{K-k} \boldsymbol{B} \boldsymbol{v}_k. \tag{6.112}$$

Premultiplying with $\boldsymbol{P}$ gives

$$\boldsymbol{P}\boldsymbol{x} = \sum_{k=1}^{K} \begin{bmatrix} \boldsymbol{J}^{K-k} & \boldsymbol{0} \\ \boldsymbol{0} & \boldsymbol{N}^{K-k} \end{bmatrix} \boldsymbol{P}\boldsymbol{B}\boldsymbol{v}_k. \tag{6.113}$$

Let $\boldsymbol{P}^{(J)}$ be the submatrix of $\boldsymbol{P}$ formed using its first J rows. Thus, we arrive at

$$\boldsymbol{P}^{(J)}\boldsymbol{x} = \sum_{k=1}^{K} \boldsymbol{J}^{K-k} \boldsymbol{P}^{(J)} \boldsymbol{B} \boldsymbol{v}_k. \tag{6.114}$$

Therefore, the system $(\boldsymbol{J}, \boldsymbol{P}^{(J)}\boldsymbol{B})$ is nonnegative controllable. By Proposition 6.3, $(\boldsymbol{J}, \boldsymbol{P}^{(J)}\boldsymbol{B})$ is s-sparse nonnegative controllable. As a result, for any $z^{(1)} \in \mathbb{R}^J$, there exist an integer $K < \infty$ and inputs $\{\boldsymbol{u}_k \in \mathbb{R}^m\}_{k=1}^K$ such that

$$z^{(1)} = \sum_{k=1}^{K} \boldsymbol{J}^{K-k} \boldsymbol{P}^{(J)} \boldsymbol{B} \boldsymbol{u}_k. \tag{6.115}$$

Therefore, we get for any $z = \begin{bmatrix} z^{(1)} \\ z^{(2)} \end{bmatrix} \in \mathbb{R}^n$,

$$
\begin{aligned}
P^{-1} \begin{bmatrix} J^n & 0 \\ 0 & 0 \end{bmatrix} z = P^{-1} \begin{bmatrix} J^n & 0 \\ 0 & 0 \end{bmatrix} \begin{bmatrix} z^{(1)} \\ z^{(2)} \end{bmatrix} = P^{-1} \begin{bmatrix} J^n z^{(1)} \\ 0 \end{bmatrix} && (6.116) \\
= P^{-1} \begin{bmatrix} J^n \sum_{k=1}^{K} J^{K-k} P^{(J)} B u_k \\ 0 \end{bmatrix} && (6.117) \\
= \sum_{k=1}^{K} P^{-1} \begin{bmatrix} J^{n+K-k} & 0 \\ 0 & 0 \end{bmatrix} P B u_k && (6.118) \\
= P^{-1} \sum_{k=1}^{K} A^{n+K-k} B u_k, && (6.119)
\end{aligned}
$$

where we use (6.111). Finally, since (6.111) also implies that

$$
\text{ColSpace}\left\{A^n\right\} = \text{ColSpace}\left\{P^{-1} \begin{bmatrix} J^n & 0 \\ 0 & 0 \end{bmatrix}\right\}, \tag{6.120}
$$

and we arrive at the desired result. □

Proof (of Theorem 6.7) The necessity of the condition is straightforward because s-sparse nonnegative controllability is more stringent than nonnegative controllability and s-sparse controllability.

To prove the sufficiency, we suppose that the conditions in Theorem 6.7 hold. So, the sparsity level $s \geq n - \text{rank}\{A\}$. Then, by Lemma 6.4, for any $z \in \mathbb{R}^p$ and $i = 0, 1, \ldots, n-1$, there exists a vector $u_{i+1} \in \mathbb{R}_+^n$ with $\|u_{i+1}\|_0 \leq s$ such that

$$
\begin{aligned}
A^i \left(A^i\right)^\dagger z = \left[A^i \left(A^i\right)^\dagger - A^{i+1} \left(A^{i+1}\right)^\dagger\right] z + A^{i+1} \left(A^{i+1}\right)^\dagger z && (6.121) \\
= \left[I - A^{i+1} \left(A^{i+1}\right)^\dagger\right] A^i B u_{i+1} + A^{i+1} \left(A^{i+1}\right)^\dagger z && (6.122) \\
= A^i B u_{i+1} + A^{i+1} \left(A^{i+1}\right)^\dagger \left[z - A^i B u_{i+1}\right]. && (6.123)
\end{aligned}
$$

Therefore, recursively using (6.123) for $i = 0, 1, \ldots, n-1$, we get for any $y \in \mathbb{R}^n$,

$$
\begin{aligned}
y = B u_1 + A A^\dagger \left[y - B u_1\right] && (6.124) \\
= B u_1 + A B u_2 + A^2 \left(A^2\right)^\dagger \left[y - B u_1 - A B u_2\right] && (6.125) \\
= \sum_{k=1}^{n} A^{k-1} B u_k + A^n \left(A^n\right)^\dagger \left[y - \sum_{k=1}^{n} A^{k-1} B u_k\right], && (6.126)
\end{aligned}
$$

where we substitute with $z = y - \sum_{k=1}^{l} A^{k-1} B u_k$ in (6.123) for $l = 0, 1, \ldots, n-1$, and $u_k \in \mathbb{R}_+^n$ has $\|u_k\|_0 \leq s$, for $k = 1, 2, \ldots, n$.

Furthermore, the last term on the right-hand side of (6.126) belongs to the set ColSpace $\left\{\boldsymbol{A}^n\right\}$. So, by Lemma 6.5, there exist an integer $K < \infty$ and input $\{\boldsymbol{u}_k \in \mathbb{R}^m\}_{k=n+1}^{n+K}$ such that

$$\boldsymbol{A}^n \left(\boldsymbol{A}^n\right)^{\dagger} \left[\boldsymbol{y} - \sum_{k=1}^{n} \boldsymbol{A}^{k-1} \boldsymbol{B} \boldsymbol{u}_k\right] = \sum_{k=1}^{K} \boldsymbol{A}^{n+K-k} \boldsymbol{B} \boldsymbol{u}_k. \tag{6.127}$$

Combining (6.126) and (6.127), we prove the desired result. □

We can restate the above theorem using Theorem 4.6 and Theorem 6.6 as follows:

Corollary 6.1 *For any sparsity level,* $s \geq 1$, *the linear dynamical system* ($\boldsymbol{A} \in \mathbb{R}^{n\times n}$, $\boldsymbol{B} \in \mathbb{R}^{n\times m}$) *in* (6.1) *is* s*-sparse nonnegative controllable if and only if the following conditions hold:*

1. *There exists no tuple* $(\lambda \in \mathbb{C}, z \neq \boldsymbol{0})$ *such that* $z^{\mathsf{T}}\boldsymbol{A} = \lambda z^{\mathsf{T}}$ *and* $z^{\mathsf{T}}\boldsymbol{B} = \boldsymbol{0}$.
2. *There exists no tuple* $(\lambda \geq 0, z \neq \boldsymbol{0})$ *such that* $z^{\mathsf{T}}\boldsymbol{A} = \lambda z^{\mathsf{T}}$ *and* $z^{\mathsf{T}}\boldsymbol{B} \in \mathbb{R}^m_+$.
3. *The sparsity level* $s \geq n - \operatorname{rank}\{\boldsymbol{A}\}$.

The above result implies that if the system $(\boldsymbol{A}, \boldsymbol{B})$ is s-sparse nonnegative controllable, it is also s-sparse nonpositive controllable. This is because if the system $(\boldsymbol{A}, \boldsymbol{B})$ satisfies the conditions of Theorem 6.7, the system $(\boldsymbol{A}, -\boldsymbol{B})$ also satisfies them.

We close this section by highlighting the main difference between sparse controllability and sparse nonnegative controllability. Sparse nonnegative controllability does not ensure that there exists a common actuator schedule that works for any initial and final states, as guaranteed by Theorem 4.4 for sparse controllability. Consequently, to design sparse nonnegative control to reach a desired final state, based on the above results, we need to resort to the state-dependent actuator scheduling approach in Section 5.2. We skip the details of the algorithms in this book.

6.4 Summary

- All stabilizable linear dynamical systems are s-sparse stabilizable, for any sparsity level $s \geq 1$.
- If a linear dynamical system is s-sparse output controllable, there exists an actuator schedule independent of the initial and final states, which can guarantee the output controllability of the system.
- The minimum sparsity level $s = s^*$ that guarantees s-sparse output controllability of an output controllable system $(\boldsymbol{A}, \boldsymbol{B}, \boldsymbol{C})$ is

$$\max_{1\leq i\leq n} \frac{1}{i}\left(p - \operatorname{rank}\left\{\boldsymbol{C}\boldsymbol{A}^i\boldsymbol{\Phi}\right\}\right) \leq s^* \leq \max_{0\leq i\leq n-1} \operatorname{rank}\left\{\boldsymbol{C}\boldsymbol{A}^i\boldsymbol{\Phi}\right\} - \operatorname{rank}\left\{\boldsymbol{C}\boldsymbol{A}^{i+1}\boldsymbol{\Phi}\right\},$$

  where $\boldsymbol{C} \in \mathbb{R}^{p\times n}$ and $\boldsymbol{\Phi}$ is the controllability matrix.

- A linear dynamical system is s-sparse nonnegative controllable if and only if it is nonnegative controllable and s-sparse controllable.

Notes

The proof of the classical stabilizability result can be found in Hespanha (2018 Chapter 8). The results on sparse stabilizability are from Sriram et al. (2022). This paper also discusses other variants of stabilizability and algorithms to design sparse control inputs to stabilize the system. The results on sparse output controllability are from Joseph (2022). The paper also presents some insightful examples and discussion. The result on nonnegative controllability is from Yoshida et al. (1994). The sparse nonnegative controllability-related results are from Joseph et al. (2021). Finally, some compressed sensing algorithms to design nonnegative sparse control can be found in Wu et al. (2014); Nguyen et al. (2019); Bruckstein et al. (2008) and their references and cited articles.

References

Bruckstein AM, Elad M, Zibulevsky M (2008) On the uniqueness of nonnegative sparse solutions to underdetermined systems of equations. IEEE Transactions on Information Theory 54(11):4813–4820

Hespanha JP (2018) Linear systems theory. Princeton University Press

Joseph G (2022) Output controllability of a linear dynamical system with sparse controls. IEEE Transactions on Control of Network Systems 10(1):147–156

Joseph G, Nettasinghe B, Krishnamurthy V, Varshney PK (2021) Controllability of network opinion in erdös-rènyi graphs using sparse control inputs. SIAM journal on control and optimization 59(3):2321–2345

Nguyen TT, Idier J, Soussen C, Djermoune EH (2019) Non-negative orthogonal greedy algorithms. IEEE Transactions on Signal Processing 67(21):5643–5658

Sriram C, Joseph G, Murthy CR (2022) Stabilizability of linear dynamical systems using sparse control inputs. IEEE Transactions on Automatic Control

Wu L, Yang Y, Liu H (2014) Nonnegative-lasso and application in index tracking. Computational Statistics & Data Analysis 70:116–126

Yoshida H, Tanada Y, Tanaka T, Yunokuchi K (1994) Controllability of multiple input discrete-time linear systems with positive controls. Transactions of the Society of Instrument and Control Engineers 30(3):243–252

Appendix A
Toolbox and Proofs

A.1 Useful Linear Algebra Identities

In this section, we present some linear algebra identities used in this book. Additional information and proofs are available in several linear algebra textbooks, like Horn and Johnson (2012); Simovici (2012); David (2011); Golub and Van Loan (2013); Strang (2012)

Theorem A.1 (Sylvester rank inequality) *For any integers $m, n, k > 0$, the matrices $\boldsymbol{A} \in \mathbb{R}^{m\times n}$ and $\boldsymbol{B} \in \mathbb{R}^{n\times k}$ satisfies*

$$\text{rank}\left\{\boldsymbol{A}\boldsymbol{B}\right\} \geq \text{rank}\left\{\boldsymbol{A}\right\} + \text{rank}\left\{\boldsymbol{B}\right\} - n. \tag{A.1}$$

Theorem A.2 (Woodbury matrix identity) *Any conformable matrices $\boldsymbol{A}$, $\boldsymbol{U}$, $\boldsymbol{C}$ and $\boldsymbol{V}$ satisfy*

$$(\boldsymbol{A} + \boldsymbol{U}\boldsymbol{C}\boldsymbol{V})^{-1} = \boldsymbol{A}^{-1} - \boldsymbol{A}^{-1}\boldsymbol{U}\left(\boldsymbol{C}^{-1} + \boldsymbol{V}\boldsymbol{A}^{-1}\boldsymbol{U}\right)^{-1}\boldsymbol{V}\boldsymbol{A}^{-1}, \tag{A.2}$$

Theorem A.3 (Weinstein–Aronszajn identity) *For any integers $m, n > 0$, the matrices $\boldsymbol{A} \in \mathbb{R}^{m\times n}$ and $\boldsymbol{B} \in \mathbb{R}^{n\times m}$ satisfies*

$$\left|\boldsymbol{I} + \boldsymbol{A}\boldsymbol{B}\right| = \left|\boldsymbol{I} + \boldsymbol{B}\boldsymbol{A}\right|. \tag{A.3}$$

A.2 Derivation of Sparse Bayesian Learning

This section details the derivation of the sparse Bayesian learning update equation (2.25) discussed in Section 2.3.4.

Consider the general compressive sensing problem that aims to reconstruct a sparse vector $\boldsymbol{x} \in \mathbb{R}^N$ from a noisy linear measurement $\boldsymbol{y} \in \mathbb{R}^M$ given by

G. Joseph and C. R. Murthy, *Sparsity-Constrained Linear Dynamical Systems*, Springer Tracts in Electrical and Electronics Engineering,
https://doi.org/10.1007/978-981-97-7090-8

$$\boldsymbol{y} = \boldsymbol{M}\boldsymbol{x} + \boldsymbol{w}, \tag{A.4}$$

where $\boldsymbol{M} \in \mathbb{R}^{M \times N}$ is the measurement matrix and $\boldsymbol{w} \in \mathbb{R}^M$ is the measurement noise. We assume that $\boldsymbol{w} \sim \mathcal{N}(\boldsymbol{0}, \sigma^2 \boldsymbol{I})$, and for simplicity, we also assume that $\sigma^2 > 0$ is known.[1] So, we have

$$p(\boldsymbol{w}; \sigma^2) = \frac{1}{(2\pi\sigma^2)^{M/2}} \exp\left[-\frac{\|\boldsymbol{w}\|^2}{2\sigma^2}\right]. \tag{A.5}$$

To derive the algorithm, we choose a fictitious Gaussian prior on $\boldsymbol{x}$ with zero mean and an unknown diagonal covariance matrix, $\boldsymbol{\Gamma} \in \mathbb{R}^{N \times N}$,

$$\boldsymbol{\Gamma} = \begin{bmatrix} \gamma_1 & & & \\ & \gamma_2 & & \\ & & \ddots & \\ & & & \gamma_n \end{bmatrix} = \text{diag}\left\{\boldsymbol{\gamma}\right\}. \tag{A.6}$$

Here, $\boldsymbol{\gamma} \in \mathbb{R}_+^N$ is the known hyperparameter vector. As a result, we arrive at

$$p(\boldsymbol{x}; \boldsymbol{\gamma}) = \frac{1}{\sqrt{|2\pi \, \text{diag}\left\{\boldsymbol{\gamma}\right\}|}} \exp\left[-\frac{1}{2}\boldsymbol{x}^\mathsf{T} \, \text{diag}\left\{\boldsymbol{\gamma}\right\}^{-1} \boldsymbol{x}\right]. \tag{A.7}$$

With the above prior distributions, we estimate the hyperparameters $\boldsymbol{\gamma}$ using $\boldsymbol{y}$ by maximizing the likelihood of the measurement. Since $\boldsymbol{x}$ and $\boldsymbol{w}$ are Gaussian distributed, the measurement is also Gaussian distributed with mean zero and covariance matrix $\boldsymbol{M} \, \text{diag}\left\{\boldsymbol{\gamma}\right\} \boldsymbol{M}^\mathsf{T} + \sigma^2 \boldsymbol{I}$. As a result, the likelihood of the measurement is given by

$$p(\boldsymbol{y}|\boldsymbol{\gamma}; \sigma^2) = \frac{1}{\sqrt{\left|2\pi \left(\boldsymbol{M} \, \text{diag}\left\{\boldsymbol{\gamma}\right\} \boldsymbol{M}^\mathsf{T} + \sigma^2 \boldsymbol{I}\right)\right|}} \times \exp\left[-\frac{1}{2}\boldsymbol{y}^\mathsf{T} \left(\boldsymbol{M} \, \text{diag}\left\{\boldsymbol{\gamma}\right\} \boldsymbol{M}^\mathsf{T} + \sigma^2 \boldsymbol{I}\right)^{-1} \boldsymbol{y}\right]. \tag{A.8}$$

The ML estimate $\boldsymbol{\gamma}_{\text{ML}}$ of $\boldsymbol{\gamma}$ is obtained by minimizing the negative log likelihood, $\boldsymbol{\gamma}_{\text{ML}} = \underset{\boldsymbol{\gamma} \in \mathbb{R}_+^N}{\arg\min} \; -\log p(\boldsymbol{y}|\boldsymbol{\gamma}; \sigma^2)$. On simplification, the optimization problem reduces to

$$\boldsymbol{\gamma}_{\text{ML}} = \underset{\boldsymbol{\gamma} \in \mathbb{R}_+^N}{\arg\min} \log\left|\boldsymbol{M} \, \text{diag}\left\{\boldsymbol{\gamma}\right\} \boldsymbol{M}^\mathsf{T} + \sigma^2 \boldsymbol{I}\right| + \boldsymbol{y}^\mathsf{T} \left(\boldsymbol{M} \, \text{diag}\left\{\boldsymbol{\gamma}\right\} \boldsymbol{M}^\mathsf{T} + \sigma^2 \boldsymbol{I}\right)^{-1} \boldsymbol{y}. \tag{A.9}$$

---

[1] There are other versions of sparse Bayesian learning where σ^2 is unknown Wipf and Rao (2004).

Unfortunately, the above optimization problem does not have a closed-form solution, and we resort to an iterative solution via the expectation–maximization algorithm. Every iteration of the expectation–maximization algorithm finds the estimate of $\boldsymbol{\gamma}$ that maximizes the expected value of the log-likelihood function of $\boldsymbol{\gamma}$, given by $\log p(\boldsymbol{x}, \boldsymbol{y}; \boldsymbol{\gamma}, \sigma^2)$, with respect to the current conditional distribution of $\boldsymbol{x}$ given $\boldsymbol{y}$ and the current estimates of the parameter $\boldsymbol{\gamma}'$. Therefore, an iteration of the expectation–maximization algorithm can be summarized as

$$\boldsymbol{\gamma}^* = \arg\max_{\boldsymbol{\gamma}\in\mathbb{R}_+^N} \mathbb{E}_{\boldsymbol{x}|\boldsymbol{y};\boldsymbol{\gamma}',\sigma^2} \left\{\log p(\boldsymbol{x}, \boldsymbol{y}; \operatorname{diag}\{\boldsymbol{\gamma}\}, \sigma^2)\right\}. \tag{A.10}$$

We simplify the above optimization as follows:

$$\log p(\boldsymbol{x}, \boldsymbol{y}; \operatorname{diag}\{\boldsymbol{\gamma}\}, \sigma^2) = \log p(\boldsymbol{y}|\boldsymbol{x}; \operatorname{diag}\{\boldsymbol{\gamma}\}, \sigma^2) + \log p(\boldsymbol{x}; \operatorname{diag}\{\boldsymbol{\gamma}\}, \sigma^2). \tag{A.11}$$

Using (A.5), we obtain

$$p(\boldsymbol{y}|\boldsymbol{x}; \operatorname{diag}\{\boldsymbol{\gamma}\}, \sigma^2) = p(\boldsymbol{y}|\boldsymbol{x}; \sigma^2) = \frac{1}{(2\pi\sigma^2)^{M/2}} \exp\left[\frac{-\|\boldsymbol{y} - \boldsymbol{M}\boldsymbol{x}\|^2}{2\sigma^2}\right], \tag{A.12}$$

which is independent of $\boldsymbol{\gamma}$. So, from (A.7) and (A.11), the optimization problem in (A.10) simplifies to

$$\boldsymbol{\gamma}^* = \arg\max_{\boldsymbol{\gamma}\in\mathbb{R}_+^N} \mathbb{E}_{\boldsymbol{x}|\boldsymbol{y};\boldsymbol{\gamma}',\sigma^2} \{\log p(\boldsymbol{x}; \operatorname{diag}\{\boldsymbol{\gamma}\})\} \tag{A.13}$$

$$= \arg\max_{\boldsymbol{\gamma}\in\mathbb{R}_+^N} \mathbb{E}_{\boldsymbol{x}|\boldsymbol{y};\boldsymbol{\gamma}',\sigma^2} \left\{\log \frac{1}{\sqrt{2\pi}\,|\operatorname{diag}\{\boldsymbol{\gamma}\}|} \exp\left[-\frac{1}{2}\boldsymbol{x}^\mathsf{T} \operatorname{diag}\{\boldsymbol{\gamma}\}^{-1} \boldsymbol{x}\right]\right\} \tag{A.14}$$

$$= \arg\min_{\boldsymbol{\gamma}\in\mathbb{R}_+^N} \mathbb{E}_{\boldsymbol{x}|\boldsymbol{y};\boldsymbol{\gamma}',\sigma^2} \left\{\log |\operatorname{diag}\{\boldsymbol{\gamma}\}| + \boldsymbol{x}^\mathsf{T} \operatorname{diag}\{\boldsymbol{\gamma}\}^{-1} \boldsymbol{x}\right\} \tag{A.15}$$

$$= \arg\min_{\boldsymbol{\gamma}\in\mathbb{R}_+^N} \log |\operatorname{diag}\{\boldsymbol{\gamma}\}| + \operatorname{tr}\left\{\operatorname{diag}\{\boldsymbol{\gamma}\}^{-1} \mathbb{E}_{\boldsymbol{x}|\boldsymbol{y};\boldsymbol{\gamma}',\sigma^2} \left\{\boldsymbol{x}\boldsymbol{x}^\mathsf{T}\right\}\right\} \tag{A.16}$$

$$= \arg\min_{\boldsymbol{\gamma}\in\mathbb{R}_+^N} \sum_{i=1}^{N} \log \boldsymbol{\gamma}_i + \frac{\mathbb{E}_{\boldsymbol{x}|\boldsymbol{y};\boldsymbol{\gamma}',\sigma^2} \left\{\boldsymbol{x}_i^2\right\}}{\boldsymbol{\gamma}_i}. \tag{A.17}$$

We notice that the above optimization problem is separable in the entries of $\boldsymbol{\gamma}$, and therefore, the optimal solution is

$$\boldsymbol{\gamma}_i^* = \arg\min_{\boldsymbol{\gamma}_i\in\mathbb{R}_+} \log \boldsymbol{\gamma}_i + \frac{\mathbb{E}_{\boldsymbol{x}|\boldsymbol{y};\boldsymbol{\gamma}',\sigma^2} \left\{\boldsymbol{x}_i^2\right\}}{\boldsymbol{\gamma}_i} = \mathbb{E}_{\boldsymbol{x}|\boldsymbol{y};\boldsymbol{\gamma}',\sigma^2} \left\{\boldsymbol{x}_i^2\right\}, \tag{A.18}$$

where the last step is obtained by setting the gradient of the convex objective function to zero. To complete the derivation, we next obtain the distribution $p(\boldsymbol{x}|\boldsymbol{y};\boldsymbol{\gamma}',\sigma^2)$. To this end, we start with the Bayes rule to derive

$$p(\boldsymbol{x}|\boldsymbol{y};\boldsymbol{\gamma}',\sigma^2)=\frac{p(\boldsymbol{y}|\boldsymbol{x};\boldsymbol{\gamma}',\sigma^2)p(\boldsymbol{x};\boldsymbol{\gamma}',\sigma^2)}{p(\boldsymbol{y};\boldsymbol{\gamma}',\sigma^2)}. \tag{A.19}$$

Substituting from (A.7), (A.8), and (A.12) with $\boldsymbol{\Gamma}'=\operatorname{diag}\{\boldsymbol{\gamma}'\}$, leads to

$$\begin{aligned}
&p(\boldsymbol{x}|\boldsymbol{y};\boldsymbol{\gamma}',\sigma^2)\\
&=\frac{1}{(2\pi\sigma^2)^{M/2}}\exp\left[-\frac{\|\boldsymbol{y}-\boldsymbol{M}\boldsymbol{x}\|^2}{2\sigma^2}\right]\times\frac{1}{\sqrt{|2\pi\boldsymbol{\Gamma}'|}}\exp\left[-\frac{\boldsymbol{x}^\top\boldsymbol{\Gamma}'^{-1}\boldsymbol{x}}{2}\right]\\
&\times\left[\frac{1}{\sqrt{|2\pi\boldsymbol{M}\boldsymbol{\Gamma}'\boldsymbol{M}^\top+\sigma^2\boldsymbol{I}|}}\exp\left[-\frac{1}{2}\boldsymbol{y}^\top\left(\boldsymbol{M}\boldsymbol{\Gamma}'\boldsymbol{M}^\top+\sigma^2\boldsymbol{I}\right)^{-1}\boldsymbol{y}\right]\right]^{-1}\\
&=\frac{1}{(2\pi)^{N/2}\sqrt{(\sigma^2)^M\,|\boldsymbol{\Gamma}'|\,|\boldsymbol{M}\boldsymbol{\Gamma}'\boldsymbol{M}^\top+\sigma^2\boldsymbol{I}|^{-1}}}\\
&\times\exp\left[-\frac{\|\boldsymbol{y}-\boldsymbol{M}\boldsymbol{x}\|^2}{2\sigma^2}-\frac{\boldsymbol{x}^\top\boldsymbol{\Gamma}'^{-1}\boldsymbol{x}}{2}+\frac{1}{2}\boldsymbol{y}^\top\left(\boldsymbol{M}\boldsymbol{\Gamma}'\boldsymbol{M}^\top+\sigma^2\boldsymbol{I}\right)^{-1}\boldsymbol{y}\right].
\end{aligned} \tag{A.20}$$

Invoking the Woodbury matrix identity (see Theorem A.2), the exponent term in (A.20) simplifies as

$$\begin{aligned}
&-\frac{\|\boldsymbol{y}-\boldsymbol{M}\boldsymbol{x}\|^2}{2\sigma^2}-\frac{\boldsymbol{x}^\top\boldsymbol{\Gamma}'^{-1}\boldsymbol{x}}{2}+\frac{1}{2}\boldsymbol{y}^\top\left(\boldsymbol{M}\boldsymbol{\Gamma}'\boldsymbol{M}^\top+\sigma^2\boldsymbol{I}\right)^{-1}\boldsymbol{y}\\
&\quad=-\frac{1}{2\sigma^2}\|\boldsymbol{y}\|^2+\frac{1}{\sigma^2}\boldsymbol{y}^\top\boldsymbol{M}\boldsymbol{x}-\frac{1}{2}\boldsymbol{x}^\top\left(\frac{1}{\sigma^2}\boldsymbol{M}^\top\boldsymbol{M}+\boldsymbol{\Gamma}'^{-1}\right)\boldsymbol{x}\\
&\qquad+\frac{1}{2}\boldsymbol{y}^\top\left[\sigma^{-2}\boldsymbol{I}-\sigma^{-4}\boldsymbol{M}\left(\boldsymbol{\Gamma}'^{-1}+\sigma^{-2}\boldsymbol{M}^\top\boldsymbol{M}\right)^{-1}\boldsymbol{M}^\top\right]\boldsymbol{y}
\end{aligned} \tag{A.21}$$

$$\begin{aligned}
&\quad=-\frac{1}{2\sigma^2}\|\boldsymbol{y}\|^2+\frac{1}{\sigma^2}\boldsymbol{y}^\top\boldsymbol{M}\boldsymbol{x}-\frac{1}{2}\boldsymbol{x}^\top\left(\frac{1}{\sigma^2}\boldsymbol{M}^\top\boldsymbol{M}+\boldsymbol{\Gamma}'^{-1}\right)\boldsymbol{x}\\
&\qquad+\frac{1}{2\sigma^2}\|\boldsymbol{y}\|^2-\frac{1}{2\sigma^4}\boldsymbol{y}^\top\boldsymbol{M}\left(\frac{1}{\sigma^2}\boldsymbol{M}^\top\boldsymbol{M}+\boldsymbol{\Gamma}'^{-1}\right)^{-1}\boldsymbol{M}^\top\boldsymbol{y}
\end{aligned} \tag{A.22}$$

$$=\boldsymbol{\mu}'^\top\boldsymbol{\Sigma}'^{-1}\boldsymbol{x}-\frac{1}{2}\boldsymbol{x}^\top\boldsymbol{\Sigma}'^{-1}\boldsymbol{x}-\frac{1}{2}\boldsymbol{\mu}'^\top\boldsymbol{\Sigma}'^{-1}\boldsymbol{\mu}' \tag{A.23}$$

$$= -\frac{1}{2}\left(x - \mu'\right) \Sigma^{-1} \left(x - \mu'\right), \tag{A.24}$$

where we define

$$\Sigma' = \left(\Gamma'^{-1} + \sigma^{-2} M^{\mathsf{T}} M\right)^{-1} \tag{A.25}$$

$$\mu' = \sigma^{-2}\left(\Gamma'^{-1} + \sigma^{-2} M^{\mathsf{T}} M\right)^{-1} M^{\mathsf{T}} y = \sigma^{-2} \Sigma' M^{\mathsf{T}} y. \tag{A.26}$$

Similarly, the determinant term in (A.24) simplifies to

$$(\sigma^2)^M \left|\Gamma'\right| \left|M\Gamma' M^{\mathsf{T}} + \sigma^2 I\right|^{-1} = \left|\Gamma'\right| \left|\sigma^{-2} M\Gamma' M^{\mathsf{T}} + I\right|^{-1} \tag{A.27}$$

$$= \left|\Gamma'\right| \left|\sigma^{-2}\Gamma' M^{\mathsf{T}} M + I\right|^{-1} \tag{A.28}$$

$$= \left|\sigma^{-2} M^{\mathsf{T}} M + \Gamma'^{-1}\right|^{-1} = \left|\Sigma'\right|, \tag{A.29}$$

where (A.28) uses the Weinstein–Aronszajn identity (see Theorem A.3). Finally, combining (A.20), (A.24), and (A.24), we obtain

$$p(x|y; \gamma', \sigma^2) = \frac{1}{\sqrt{(2\pi)^N \left|\Sigma'\right|}} \exp\left[-\frac{1}{2}(x - \mu')^{\mathsf{T}} \Sigma'^{-1} (x - \mu')\right]. \tag{A.30}$$

Hence, (A.18) reduces to $\gamma_i^* = \mu_i'^2 + \Sigma'_{ii}$, establishing (2.25), and the derivation of the sparse Bayesian learning algorithm is complete.

A.3 Probability and Statistics Toolbox

We present some results that are used to prove Theorem 3.5.

A.3.1 Sub-Gaussian and Subexponential Random Variables

We start with the definition of sub-Gaussian matrices.

Definition A.1 A random variable Z is said to be sub-Gaussian with parameters $\beta, \kappa > 0$ if, for any $t > 0$, it satisfies

$$\mathbb{P}\{|Z| \geq t\} \leq \beta \exp(-\kappa t^2). \tag{A.31}$$

The sub-Gaussian distribution has strong tail decay so that the tails are exponentially bounded. Some example distributions are Gaussian, Rademacher, uniform distribution over a bounded interval, etc. It is easy to see that any bounded random variable is sub-Gaussian.

Sub-Gaussian random variables are closely related to another class of random variables called subexponential random variables.

Definition A.2 A random variable Z is said to be subexponential with parameter $\beta, \kappa > 0$ if, for any $t > 0$, it satisfies

$$\mathbb{P}\{|Z| \geq t\} \leq \beta \exp(-\kappa t). \tag{A.32}$$

The following lemma relates sub-Gaussian and subexponential random variables.

Lemma A.1 *A random variable Z is sub-Gaussian if and only if Z^2 is subexponential.*

Proof The result is straightforward from Definitions A.1 and A.2. □

We know that sum of scaled Gaussian random variables is another Gaussian random variable. A similar property holds for zero-mean sub-Gaussian random variables as well.

Lemma A.2 (Foucart and Rauhut (2013, Proposition 7.24, Theorem 7.27.)) *Let $Z_1, Z_2, \ldots, Z_M$ be M independent zero-mean sub-Gaussian random variables with parameters β, κ in Definition A.1. Then, for any $\boldsymbol{a} \in \mathbb{R}^M$, the random variable $\sum_{i=1}^{M} \boldsymbol{a}_i Z_i$ is a zero-mean sub-Gaussian with parameters $\beta, \kappa / \|\boldsymbol{a}\|^2$.*

A.3.2 Concentration Inequalities for Subexponential Random Variables

We start with the famous Bernstein inequality for zero-mean subexponential random variables without proof.

Proposition A.1 (Foucart and Rauhut 2013, Corollary 7.32) *Let $Z_1, Z_2, \ldots, Z_M$ be independent zero-mean subexponential random variables such that $\mathbb{P}\{|Z_i| > t\} \leq \beta \exp(-\kappa t)$. Then, for any $t > 0$, they satisfy*

$$\mathbb{P}\left\{\sum_{i=1}^{M} Z_i > t\right\} \leq \exp\left(-\frac{\kappa^2 t^2/2}{2\beta M + \kappa t}\right). \tag{A.33}$$

Using the above theorem, we can derive a similar result for non-centered subexponential random variables.

Lemma A.3 *Let $Z_1, Z_2, \ldots, Z_M$ be independent subexponential random variables with $\mu_{\min} \leq \mathbb{E}(Z_i) \leq \mu_{\max}$, for $i = 1, 2, \ldots, M$. Then, for any $\tau > M\tilde{\mu}$ and $\tilde{\mu} = \max\{\mu_{\max}, -\mu_{\min}\}$, they satisfy*

$$\mathbb{P}\left\{\sum_{i=1}^{M} Z_i > t\right\} \leq 2\exp\left(-\frac{\kappa^2(\tau - M\tilde{\mu})^2/2}{2\beta M + \kappa(t - M\tilde{\mu})}\right). \tag{A.34}$$

Proof From Proposition A.1, we see that, for any $t > 0$,

$$\mathbb{P}\left\{\sum_{i=1}^{M} Z_i \geq t + M\tilde{\mu}\right\} \leq \mathbb{P}\left\{\sum_{i=1}^{M} Z_i \geq t + M\mu_{\max}\right\} \tag{A.35}$$

$$\leq \mathbb{P}\left\{\sum_{i=1}^{M} Z_i \geq t + \sum_{i=1}^{M} \mathbb{E}(Z_i)\right\} \tag{A.36}$$

$$= \mathbb{P}\left\{\sum_{i=1}^{M} (Z_i - \mathbb{E}(Z_i)) \geq t\right\} \tag{A.37}$$

$$\leq \exp\left(-\frac{\kappa^2 t^2/2}{2\beta M + \kappa t}\right). \tag{A.38}$$

Then, for any $\tau = t + M\tilde{\mu} > M\tilde{\mu}$,

$$\mathbb{P}\left\{\sum_{i=1}^{M} Z_i \geq \tau\right\} \leq \exp\left(-\frac{\kappa^2(\tau - M\tilde{\mu})^2/2}{2\beta M + \kappa(t - M\tilde{\mu})}\right). \tag{A.39}$$

Similarly, invoking Proposition A.1 with $-\sum_{i=1}^{M}(Z_i - \mathbb{E}(Z_i))$, for any $\tau = t + M\tilde{\mu} > M\tilde{\mu}$, we derive

$$\mathbb{P}\left\{-\sum_{i=1}^{M} Z_i \geq \tau\right\} \leq \mathbb{P}\left\{-\sum_{i=1}^{M} (Z_i - \mathbb{E}(Z_i)) \geq \tau - M\tilde{\mu}\right\} \tag{A.40}$$

$$\leq \exp\left(-\frac{\kappa^2(\tau - M\tilde{\mu})^2/2}{2\beta M + \kappa(t - M\tilde{\mu})}\right). \tag{A.41}$$

Combining (A.38) and (A.41), for $\tau > M\tilde{\mu}$,

$$\mathbb{P}\left\{\left|\sum_{i=1}^{M} Z_i\right| \geq \tau\right\} \leq 2\exp\left(-\frac{\kappa^2(\tau - M\tilde{\mu})^2/2}{2\beta M + \kappa(\tau - M\tilde{\mu})}\right). \tag{A.42}$$

□

A.3.3 Covering Number

Let $\mathcal{M}$ be a subset of $\mathbb{R}^N$. For $t > 0$, the covering number T^* is defined as the smallest integer T such that $\mathcal{M}$ that can be covered with balls centered at points from a set $\mathcal{Z}$

of T elements, i.e.,

$$\mathcal{M} \subseteq \bigcup_{z \in \mathcal{Z}} \mathcal{B}(z, \rho). \tag{A.43}$$

Here, a ball with center z and radius $\rho > 0$ is defined as

$$\mathcal{B}(z, \rho) = \{z' \in \mathbb{R}^N : \ \|z - z'\| \leq \rho\}. \tag{A.44}$$

The set $\mathcal{Z}$ is called a ρ-covering. The minimum cardinality T of $\mathcal{Z}$ is bounded via the following proposition.

Proposition A.2 *Let $\mathcal{M}$ be a subset of the unit ball,* $\{\boldsymbol{x} \in \mathbb{R}^N : \ \|\boldsymbol{x}\| \leq 1\}$ *and $\mathcal{B}(z, \rho)$ be a ball around point $z \in \mathcal{M}$ with radius $\rho > 0$,*

$$\mathcal{B}(z, \rho) = \{z' \in \mathbb{R}^N : \ \|z - z'\| \leq \rho\}. \tag{A.45}$$

Also, let $\mathcal{Z}$ be the smallest set such that the balls around these points that covers $\mathcal{M}$,

$$\mathcal{M} \subseteq \bigcup_{z \in \mathcal{Z}} \mathcal{B}(z, \rho). \tag{A.46}$$

Then, the cardinality of $\mathcal{Z}$ is bounded by

$$|\mathcal{Z}| \leq \left(1 + \frac{2}{\rho}\right)^N. \tag{A.47}$$

Proof Let $\tilde{\mathcal{Z}} \in \mathcal{M}$ be the maximal set such that any two points in $\tilde{z}, \tilde{z}' \in \tilde{\mathcal{Z}}$ satisfy

$$\left\|\tilde{z} - \tilde{z}'\right\| > \rho. \tag{A.48}$$

Suppose there exists a point $\boldsymbol{x} \in \mathcal{M}$ such that $\boldsymbol{x} \notin \mathcal{B}(z, \rho)$, for any $z \in \tilde{\mathcal{Z}}$. Then, we can add $\boldsymbol{x}$ to $\tilde{\mathcal{Z}} \in \mathcal{M}$ and any two points in $\tilde{\mathcal{Z}} \in \mathcal{M}$ still satisfy (A.48). This is a contradiction to the maximality of the set. Therefore, any point $\boldsymbol{x} \in \mathcal{M}$ satisfies $\boldsymbol{x} \in \mathcal{B}(z, \rho)$, for some $z \in \tilde{\mathcal{Z}}$, and we obtain

$$\mathcal{M} \subseteq \bigcup_{z \in \tilde{\mathcal{Z}}} \mathcal{B}(z, \rho). \tag{A.49}$$

Consequently, we arrive at

$$|\mathcal{Z}| \leq \left|\tilde{\mathcal{Z}}\right|. \tag{A.50}$$

Furthermore, (A.48) also implies that any point $\boldsymbol{x} \in \mathcal{B}(\tilde{z}, \rho/2)$ and $\tilde{z}' \in \tilde{\mathcal{Z}}$ with $\tilde{z}' \neq \tilde{z}$ satisfy

$$\|x - \tilde{z}'\| = \|(x - \tilde{z}) + (\tilde{z} - \tilde{z}')\| \tag{A.51}$$

$$\geq \|\tilde{z} - \tilde{z}'\| - \|x - \tilde{z}\| \tag{A.52}$$

$$\geq \|\tilde{z} - \tilde{z}'\| - \rho/2 \geq \rho - \rho/2 > \rho/2. \tag{A.53}$$

Hence, we conclude that the sets $\mathcal{B}(\tilde{z}', \rho/2)$ and $\mathcal{B}(\tilde{z}, \rho/2)$ are disjoint for any $\tilde{z}' \in \tilde{\mathcal{Z}}$ with $\tilde{z}' \neq \tilde{z}$. Consequently, we deduce that

$$\sum_{\tilde{z} \in \tilde{\mathcal{Z}}} \text{vol}\,(\mathcal{B}(\tilde{z}, \rho/2)) \leq \text{vol}\left(\bigcup_{\tilde{z} \in \tilde{\mathcal{Z}}} \mathcal{B}(\tilde{z}, \rho/2)\right). \tag{A.54}$$

Also, for any point in $x \in \mathcal{B}(\tilde{z}, \rho/2)$, we get

$$\|x\| \leq \|x - \tilde{z}\| + \|\tilde{z}\| = 1 + \rho/2. \tag{A.55}$$

Therefore, we derive that $\mathcal{B}(\tilde{z}, \rho/2) \subseteq \mathcal{B}(\mathbf{0}, 1 + \rho/2)$ for any $\tilde{z} \in \tilde{\mathcal{Z}}$, and

$$\bigcup_{\tilde{z} \in \tilde{\mathcal{Z}}} \mathcal{B}(\tilde{z}, \rho/2) \subset \mathcal{B}(\mathbf{0}, 1 + \rho/2). \tag{A.56}$$

From (A.54) and (A.56), we have

$$\sum_{\tilde{z} \in \tilde{\mathcal{Z}}} \text{vol}\,(\mathcal{B}(\tilde{z}, \rho/2)) \leq \text{vol}\,(\mathcal{B}(\mathbf{0}, 1 + \rho/2)), \tag{A.57}$$

where vol denotes the volume (Lebesgue measure) of a ball. Also, volume satisfies the homogeneity relation $\text{vol}\,(\mathcal{B}(z, R_B)) = R_B^N \text{vol}\,(\mathcal{B}(z', 1))$, for any $z, z' \in \mathbb{B}_N$ and R_B. So, we conclude that

$$\sum_{\tilde{z} \in \tilde{\mathcal{Z}}} (\rho/2)^N \text{vol}\,(\mathcal{B}(\mathbf{0}, 1)) \leq (1 + \rho/2)^N\, \text{vol}\,(\mathcal{B}(\mathbf{0}, 1)). \tag{A.58}$$

Therefore, we deduce

$$\left|\tilde{\mathcal{Z}}\right| (\rho/2)^N \leq (1 + \rho/2)^N. \tag{A.59}$$

Rearranging the above relation and combining it with (A.50), we arrive at the desired result. □

References

David WL (2011) Matrix Theory. Allied Publishers Pvt, Limited
Foucart S, Rauhut H (2013) An invitation to compressive sensing. Springer
Golub GH, Van Loan CF (2013) Matrix computations. JHU press
Horn RA, Johnson CR (2012) Matrix analysis. Cambridge University Press
Simovici DA (2012) Linear algebra tools for data mining. World Scientific
Strang G (2012) Linear algebra and its applications 4th ed
Wipf DP, Rao BD (2004) Sparse bayesian learning for basis selection. IEEE Transactions on Signal processing 52(8):2153–2164

Index

G. Joseph and C. R. Murthy, *Sparsity-Constrained Linear Dynamical Systems*,
Springer Tracts in Electrical and Electronics Engineering,
https://doi.org/10.1007/978-981-97-7090-8

The manufacturer's authorised representative in the EU is Springer Nature Customer Service Centre GmbH, Europaplatz 3, 69115 Heidelberg, Germany. If you have any concerns regarding our products, please contact ProductSafety@springernature.com

Printed and bound by CPI Group (UK) Ltd, Croydon, CR0 4YY

01/10/2026

02225876-0003